35
NMR
Basic Principles and Progress

Editorial Board
P. Diehl · E. Fluck · H. Günther · R. Kosfeld · J. Seelig

Springer

Berlin
Heidelberg
New York
Barcelona
Budapest
Hong Kong
London
Milan
Paris
Santa Clara
Singapore
Tokyo

Ab Initio Calculations of Conformational Effects on ^{13}C NMR Spectra of Amorphous Polymers

By R. Born and H. W. Spiess

Volume Editor: J. Seelig

 Springer

Chemistry Library

Authors

Dr. R. Born
Max-Planck-Institut für Polymerforschung
Postfach 3148
D-55021 Mainz, Germany

Prof. H. W. Spiess
Max-Planck-Institut für Polymerforschung
Postfach 3148
D-55021 Mainz, Germany
E-mail: Spiess@mpip-mainz.mpg.de

Volume Editor

Prof. J. Seelig
Abt. Biophysikalische Chemie
Biozentrum der Universität Basel
Klingelbergstraße 70
CH-4056 Basel, Switzerland
E-mail: seelig1@ubaclu.unibas.ch

ISSN 0179-5989
ISBN 3-540-62010-9
Springer-Verlag Berlin Heidelberg New York

Typeset with LATEX: Danny Lewis Book Production, Berlin
Cover: Medio GmbH, Berlin
SPIN: 10551778   66/3020 - 5 4 3 2 1 0 - Printed on acid-free paper

QD96
N8
B67
1997
CHEM

Preface

In NMR, it is well-known that the chemical shift conveys structural information, e.g. a carbonyl carbon will have a resonance frequency appreciably different from a methyl carbon, etc. The relation between structure and chemical
shift is mostly established by empirical rules on the basis of prior experience.
It is only quite recently that the advent of both comparatively cheap computing power and novel quantum chemistry approaches have provided feasible
routes to *calculate* the chemical shift at the ab initio level for molecules of
reasonable size. This raises the question whether application of these novel
theoretical concepts offers a means of obtaining new structural information
for the complex chain molecules one deals with in polymer science.

Solid state ^{13}C-NMR spectra of glassy amorphous polymers display broad,
partially structured resonance regions that reflect the underlying disorder of
the polymer chains. The chemical shift responds to the variation of the geometry of the chain, and the broad resonance regions can be explained by an
inhomogeneous superposition of various chain geometries (and thus chemical shifts). In this review, we present a novel approach to combine polymer
chain statistical models, quantum chemistry and solid state NMR to provide quantitative information about the local chain geometry in amorphous
polymers. The statistical model yields the relative occurrence of the various
geometries, and quantum chemistry (together with a force field geometry optimization) establishes the link between geometry and chemical shift. Finally,
the simulated spectra are compared with experimental ^{13}C CP/MAS spectra.
The scheme is applied to various polymer systems of different structural
classes.

The computing power of comparatively cheap work stations is still increasing rapidly, and the algorithms of quantum chemistry are becoming
more efficient as well. Using these mechanisms, the limitations imposed by
the availability of computing resources are gradually relaxed, and more sophisticated calculations are becoming possible. This naturally applies to the
calculations presented here. Even in the short period of two years between
the actual calculations and the publication of this review some computer time
limitations are already less severe. Thus, easier and more accurate calcula-

tions for the 'simple' cases considered here and meaningful simulations for more complicated structures will in the future facilitate the application of these concepts to more problems in polymer science.

The authors want to thank Dr. A. Heuer for continuous discussions on the structure of polymer glasses and for carefully reading the draft. Special thanks are to Prof. Dr. W. Kutzelnigg, Dr. M. Schindler and Dr. U. Fleischer for providing the IGLO program and for helpful discussions. Financial suport of this work by the Deutsche Forschungsgemeinschaft (SFB 262) is gratefully acknowledged.

Main, February 1997   R. Born
                                                                                              H.W. Spiess

Table of Contents

Ab Initio Calculations of Conformational Effects on ^{13}C NMR Spectra of Amorphous Polymers

R. Born and H.W. Spiess

Max-Planck-Institut für Polymerforschung, Postfach 3148, D-55021 Mainz, Germany
E-mail: Spiess@mpip-mainz.mpg.de

A new approach for the simulation of inhomogeneouly broadened solid state NMR spectra is presented. The geometry dependence of the chemical shift is investigated and used to analyze the local molecular organization of amorphous glassy polymers by comparison of simulated lineshapes with experimental spectra. The method combines statistical conformational models, force-field optimization and the advanced quantum chemical ab initio IGLO scheme for the calculation of chemical shifts. An introduction to these topics is given together with a detailed derivation of our approach. The novel scheme is used to assess the validity of statistical models of polymer conformations and to discriminate between competing theories. Moreover, the molecular orbital contributions to the chemical shift, the configurational splitting in solution, and the role of the anisotropy of the chemical shift as a source of structural information are addressed. Methodical aspects, including preconditions and limitations, are covered in a separate chapter. To check the range of applicability, the method is applied to various polymer systems of different structural classes (poly(ethylene), poly(propylene), poly(isobutylene), poly(butadiene), poly(isoprene), poly(vinyl chloride), poly(methyl methacrylate)).

Contents

NMR Basic Principles and Progress, Vol. 35
© Springer Verlag Berlin Heidelberg 1997

Abbreviations and Symbols

Important Symbols

$g_K(\sigma)$	simulated line shape for nucleus K
G^j	optimized geometry belonging to k^j
k^j	configurational/conformational sequence
p_{config}	configurational statistics
$p_{conform}^{RIS}$	conformational statistics
s_b	Gaussian broadening function
T	temperature
T_g	glass transition temperature
x_i	configuration of a diad in a pseudoasymmetric polymer, either m or r
γ_X	γ-*gauche* constant
δ	chemical shift (experimental scale), measured from TMS
δ_{aniso}	anisotropy of the chemical shift tensor
$\Delta\delta_X$	relative chemical shift
η	asymmetry of the chemical shift tensor
ϑ_i	main chain dihedral angle
σ	chemical shift (theoretical scale), with respect to the bare nucleus
σ_{xx}, σ_{yy}, σ_{zz}	principal values of the chemical shift tensor
χ_i	orientation of the COOCH$_3$ sidechain in PMMA

Abbreviations

AO	Atomic Orbital
aPP	atactic poly(propylene)
CHF	Coupled Hartree-Fock
CVFF	Consistent Valence Force Field
CP/MAS	Cross Polarization / Magic Angle Spinning
DSC	Differential Scanning Calorimetry
DZ	atomic basis of *double zeta* quality
GIAO	Gauge Including Atomic Orbitals
IGLO	Individual Gauges for Localized Molecular Orbitals

iPMMA	isotactic poly(methyl methacrylate)
iPP	isotactic poly(propylene)
LMO	Localized Molecular Orbital
LORG	Local Gauge/Local Origin
MC	Monte Carlo
MC-SCF	Multi-Configurational Self-Consistent Field
MD	Molecular Dynamics
MO	Molecular Orbital
MP	Møller-Plesset
PBD	1,4-poly(butadiene)
PE	poly(ethylene)
PIB	poly(isobutylene)
PIP	1,4-poly(isoprene)
PMMA	poly(methyl methacrylate)
PP	poly(propylene)
PVC	poly(vinyl chloride)
RIS	Rotational Isomeric State
SCF	Self-Consistent Field
sPMMA	syndiotactic poly(methyl methacrylate)
sPP	syndiotactic poly(propylene)
TMS	tetra methyl silane

1
Introduction

The way chain molecules are organized either in bulk or in solution has attracted a lot of experimental [1, 2, 3] and theoretical [4, 5, 6] interest. In this work we focus on a detailed investigation of the *local* structure of a polymer glass by a combination of solid state NMR and ab initio theory.

A typical (homo)polymer chain consists of many repetitions of the same monomeric unit (Fig. 1). In many cases, the bonds along the main chain will be σ-bonds, which introduces a considerable degree of flexibility [4, 5, 6]. The dihedral angles belonging to these bonds are not restricted to a single value, but several positions are energetically achievable, as can be seen in the simple case of butane (Fig. 2). This degree of freedom is called *conformation*. As a consequence, the pattern of molecular organization of chain molecules in bulk is much more complicated than that of simple, low molecular weight compounds. As a first possibility, the polymer melt may crystallize; in this case a single chain forms (at least for parts of its length) a helix with a regular sequence of dihedral angles. The helices themselves are in turn packed in a regular way (Fig. 3.a). These polymer crystals can be analyzed by the customary scattering methods (X-ray, neutron scattering). In principle, the positions of the nuclei, the dihedral angles and the relative positions of the helices, even the nature of the crys-

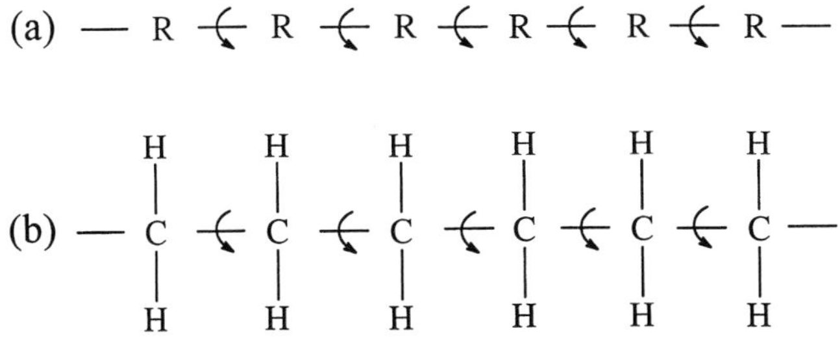

Fig. 1. A typical homopolymer consists of many repetitions of the same monomeric unit, which are connected by σ-bonds introducing a considerable degree of flexibility (**a**). Poly(ethylene) is shown as an example (**b**)

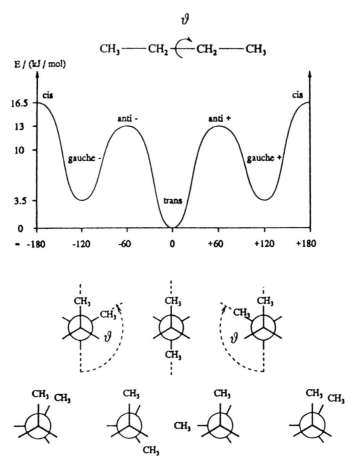

Fig. 2. The molecular energy of butane as a function of the dihedral angle ϑ is depicted (*top*). The energies of the *trans* and *gauche* states are quite low; they are separated by states of comparably high energy (*cis, anti*) which are only poorly populated themselves but still allow for rapid transitions among the low energy states. In the Newman projections (bottom), the various states are depicted (cf. Fig. 10)

talline defects [7] can be inferred from a scattering experiment, though, in practice, the open structure of polymer crystals imposes some limitations.

A second possibility of solid state molecular organization is the amorphous phase. Here the dihedral angles along the chain do not have a regular sequence of fixed values but are subject to a statistical probability distribution; the average chain forms a random Gaussian coil like in the melt or in the solution (Fig. 3b). In the amorphous case, the scattering methods provide mainly *global* parameters on a longer length scale like the radius of gyration

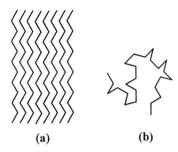

(a) (b)

Fig. 3. In bulk, polymer chains either form helices (with a regular sequence of dihedral angles along the chain) that are packed in a regular polymer crystal (**a**) or they are in a disordered amorphous state where the average chain can be described as a random Gaussian coil (**b**)

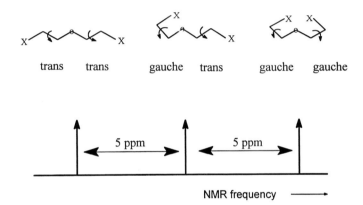

Fig. 4. The chemical shift is sensitive to the conformation (γ-*gauche* effect) of the molecule. In the extended *trans* conformation, the central carbon atom (denoted by $\circ$) experiences a chemical shift that is reduced by ≈ 5 ppm for every γ-neighbor being in *gauche* position. This simple qualitative understanding will have to be modified in the light of the ab initio calculations

(approximately the size of the random coil) and the characteristic ratio. The *local* geometry and its statistical distribution are more difficult to obtain. To investigate this topic, nuclear magnetic resonance (NMR) should be suitable because the chemical shift is susceptible not only to the chemical structure of the molecule (constitution, configuration), but also to the geometric pattern, i.e. the conformation [8]. Historically, the most important geometry dependence has long been known under the name of γ-*gauche* effect: The chemical shift of a ^{13}C nucleus is influenced by the geometrical position of its heavy atom γ-neighbors and thus by the value of the corresponding dihedral angle (Fig. 4). If a nucleus experiences a γ-neighbor in a *gauche* position, its chemical shift mostly changes by about 5 ppm with respect to the *trans*

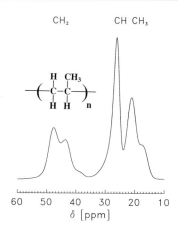

Fig. 5. Experimental spectrum of atactic poly(propylene) recorded at $T = 260$ K, below the calorimetric glass transition temperature. The various resonances are considerably broadened and partly structured

position. Thus, NMR as a local method is sensitive to the variation of the local geometry (conformation). Consequently, a thorough NMR investigation might yield structural information about the local molecular organization of the amorphous phase of polymers. In the glass, the conformational disorder is frozen on the NMR time scale and the NMR resonances are inhomogeneously broadened (Fig. 5). This reflects the presence of multiple geometries in the amorphous polymer glass. The broad, smooth character of the experimental resonances suggests that a phenomenological explanation by a few combinations of well-defined *trans-* and *gauche* states (with correspondingly sharp subresonances) is too simplistic. Instead, a continuum of geometries exists that requires a detailed investigation.

In this review, we will mainly concentrate on simulations of ^{13}C spectra in the amorphous glassy state by a novel approach, combining established microscopic statistical models and quantum chemical ab initio calculations of the chemical shift. The statistical model is used to identify the important geometries which – in a second step – are generated by an empirical forcefield and fed into an ab initio calculation of the chemical shift by the **I**ndividual **G**auges for **L**ocalized Molecular **O**rbitals (IGLO) method. Finally, the probabilities for the geometries (inferred from the statistical model) and the corresponding chemical shifts (calculated by IGLO) are combined to yield a simulated line shape (Fig. 6).

The present paper is organized as follows. In Sect. 2, a few relevant notions in polymer science (Sect. 2.1) will be introduced and discussed. Next, a short survey of the essential tools of the approach will be given, including the statistics of chain molecules (Sect. 2.2), empirical force fields (Sect. 2.3),

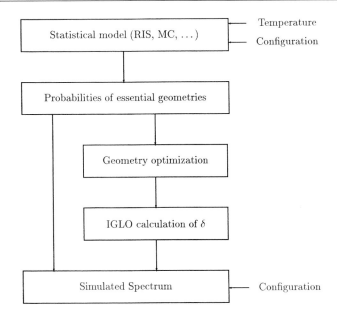

Fig. 6. Schematic outline of the ab initio method for the simulation of ^{13}C solid state NMR spectra. Important building blocks are the statistical conformational model, the technique of geometry optimization and the quantum chemical IGLO method for the computation of the chemical shifts

and the quantum chemical IGLO method for the calculation of the chemical shifts (Sect. 2.4). Finally, the geometry effects in NMR shifts are reviewed briefly (Sect. 2.5).

In the central part (Sect. 3), the simulation technique will be explained in detail and illustrated by the example of atactic poly(propylene) (Sects. 3.1–3.3). Special sections are devoted to the simulation of the solid state spectrum (Sect. 3.4), the correlation of chemical shift and geometry (Sect. 3.5), a molecular orbital (MO) analysis (Sect. 3.6), the configurational splitting in solution (Sect. 3.7) and the role of the anisotropy of the chemical shift as a source of structural information (Sect. 3.8).

In Sect. 4, some methodical aspects will be investigated, addressing the relative size of *intra*- and *inter*molecular contributions to the chemical shift (Sect. 4.1), the influence of the endgroups on the results (Sect. 4.2), the significance of the atomic basis and the empirical forcefield (Sect. 4.3), as well as some more general remarks on the preconditions of spectral simulation in disordered systems (Sect. 4.4–4.6).

In Sect. 5, it is demonstrated, that the approach systematically developed in Sect. 3 can be successfully applied to polymers of different structural classes (Table 1, Fig. 7). Our investigations include the

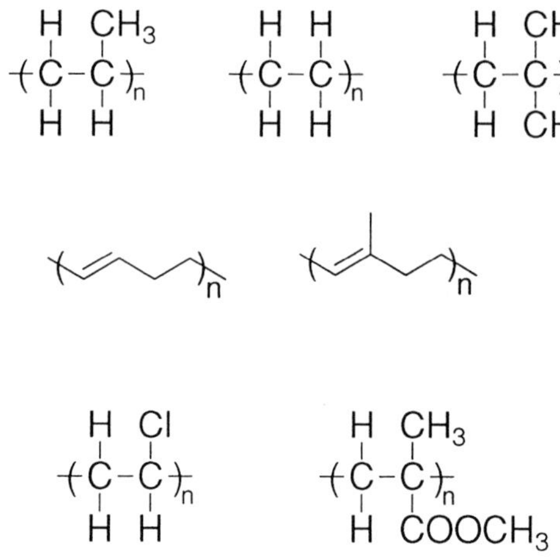

$$\begin{array}{ccc}
\text{H} \quad \text{CH}_3 & \text{H} \quad \text{H} & \text{H} \quad \text{CH}_3 \\
\text{+C-C+}_n & \text{+C-C+}_n & \text{+C-C+}_n \\
\text{H} \quad \text{H} & \text{H} \quad \text{H} & \text{H} \quad \text{CH}_3
\end{array}$$

$$\begin{array}{cc}
\text{H} \quad \text{Cl} & \text{H} \quad \text{CH}_3 \\
\text{+C-C+}_n & \text{+C-C+}_n \\
\text{H} \quad \text{H} & \text{H} \quad \text{COOCH}_3
\end{array}$$

Fig. 7. Repeat units for the investigated polymers: poly(propylene), poly(ethylene), poly(isobutylene), poly(butadiene), poly(isoprene), poly(vinyl chlorid), poly(methyl methacrylate) (from *left* to *right, top* to *bottom*)

Table 1. List of investigated polymers. The characteristic features of the system and the statistical model employed are also included. The corresponding model molecules are displayed in the respective sections

polymer	abbrev.	characteristic feature	statistical model
poly(ethylene)	PE	simplest polymer	[27]
atact. poly(propylene)	aPP	polyolefine, one sidegroup	[86]
poly(isobutylene)	PIB	polyolefine, two sidegroups	[26]
1,4-poly(butadiene)	PBD	unsaturated	[21]
1,4-poly(isoprene)	PIP	unsaturated, sidegroup	[22]
poly(vinyl chloride)	PVC	contains hetero atom	[116, 117, 118]
poly(methyl methacrylate)	PMMA	extended sidegroup	[24]

simplest polymer poly(ethylene) (Sect. 5.1), the bisubstituted polyolefine poly(isobutylene) (Sect. 5.2), the unsaturated rubbers 1,4-poly(butadiene) and 1,4-poly(isoprene) (Sect. 5.3), the hetero atom containing polymer poly(vinyl chloride) (Sect. 5.4), and poly(methyl methacrylate), with a more extended sidegroup (Sect. 5.5). In this section, mainly results will be presented, and only a few details of the methods are mentioned throughout the chapter.

To permit an easy and rapid access to special aspects, the chapters and some of the more important subsections are largely self-contained and should be understandable without extensive study of preceding chapters. Whenever possible, important arguments are shortly restated in the respective sections with a reference to the site of the more detailed discussion.

Although this paper is mainly a review, several previously unpublished results have been included, especially in Sects. 4 and 5. Shorter accounts covering special aspects have already been published. [9, 10, 11]

2
Notions and Concepts in Chain Molecules and Quantum Chemistry

2.1
Constitution, Configuration and Conformation in Amorphous Polymers

We consider a general bisubstituted vinyl polymer (Fig. 8.a) with two different sidegroups. For the characterization of its geometry, the following notions must be recognized. A more detailed account can be found in many textbooks on organic chemistry or polymer science, e.g. [8].

Constitution refers to the binding situation, i.e. the sequence of atoms in a molecule. Systems which have the same sum formula, but different structural formulae are called constitutional isomers. The transformation of one constitutional isomer into another requires breaking and forming chemical bonds with activation energies > 1 eV (96.5 kJ/mol). Polymerization defects like head-head-addition (in place of the desired head-tail-addition) introduce constitutional isomers which sometimes have pronounced effects on the material properties. For simplicity however, an ideal, defect-free constitution of the systems inquired will be assumed throughout this review, and constitutional isomerism will not be discussed.

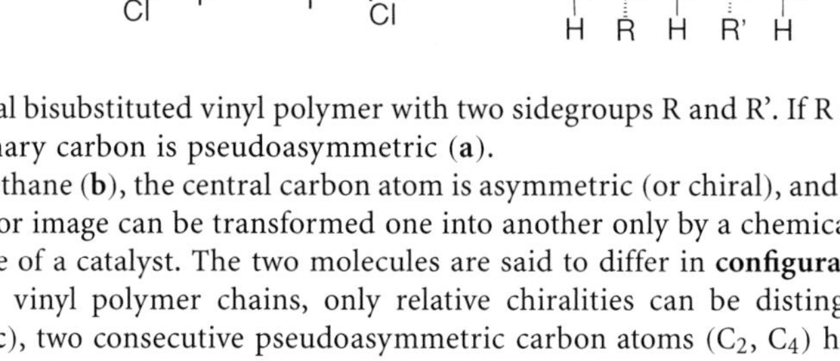

Fig. 8. Typical bisubstituted vinyl polymer with two sidegroups R and R'. If R and R' differ, the quarternary carbon is pseudoasymmetric (**a**).
In FClBr-methane (**b**), the central carbon atom is asymmetric (or chiral), and the molecule and its mirror image can be transformed one into another only by a chemical reaction or the influence of a catalyst. The two molecules are said to differ in **configuration**.
In extended vinyl polymer chains, only relative chiralities can be distinguished. In a *meso* diad (**c**), two consecutive pseudoasymmetric carbon atoms (C_2, C_4) have the same configuration, whereas in *syndiotactic* diads (**d**) the configuration is opposite

Carbon atoms with four mutually different substituents are called asymmetrically substituted (asymmetric for short), as one geometry and its mirror image are separate molecules that can be transformed one into another only by a chemical reaction typically mediated by a suitable catalyst (Fig. 8.b). The activation energy is comparable to a constitutional transformation. The **configuration** is thus a property of the molecule (fixed by the conditions of synthesis) that does not change later and normally does not depend on temperature. We do not want to dwell any longer on the complicated field of chirality in general chemistry and concentrate on the far simpler notions in vinyl polymers. The carbon atoms C_2 and C_4 of the polymer chain in Fig. 8.c ($R \neq R'$) are formally both chiral centers. But in practice, the two chain ends (denoted by the wavy line) are very similar; then only the relative configuration of C_2 with respect to C_4 is relevant, the absolute chirality being a matter of little consequence. Such carbon atoms are called pseudochiral or pseudoasymmetric. A sequence of two chiral carbon atoms in a vinyl polymer is called a *meso*-diad (*m*) if the two carbon atom C_2 and C_4 have a like configuration, and a *racemic* diad (*r*), if their configuration is opposite (Figs. 8.c/d). The configuration of a polymer segment may be thus described as a sequence of the relative configurations of the consecutive diads. A polymer with constant configuration $\cdots mmmmm \cdots$ is called *isotactic* (from the greek ἴσος, a chain with an alternating configuration $\cdots rrrrrr \cdots$ *syndiotactic* (greek σύνδυο = by twos), and a system without apparent pattern *atactic*. In a more restricted sense, only polymers with a Bernoullian configurational statistics and an equal probability for *m*- and *r*-diads are called (ideally) atactic. A *diad* is the segment between two neighboring chiral atoms (Fig. 9.a), whereas an *interdiad* is a two-bond sequence of a vinyl polymer centered around a pseudochiral carbon atom (Fig. 9.b). A single carbon atom belongs thus simultaneously to two diads and one interdiad (for substituted carbons) or to a single diad and two interdiads (for methylene carbons). A segment consisting of two consecutive diads is called a *triad*, a segment of three diads (or four chiral carbon atoms) a *tetrad* (Figs. 9.c/d), etc.

The presence of various configurations in a sample gives rise to a configurational splitting in the solution NMR spectrum. This aspect will be discussed in some more detail in Sect. 3.7.

Our main concern is *conformation*. As already mentioned in the Introduction, a dihedral angle along a σ-bond is not restricted to a single value, but various energy minima are accessible (Fig. 2). For an easy visualization, we use the so-called *Newman*-projection. For the definition of the dihedral angle ϑ_1 in Fig. 10.a we look from the bisubstituted carbon atom C_2 along the bond into the direction of the unsubstituted carbon C_1. The carbon atom C_2 is represented by the intersection of the three bonds, the atom C_1 is covered and not seen. The wavy lines denote the continuation of the chain. In contrast to the practice in ordinary organic chemistry, the dihedral angle ϑ_1 is set to 0, if the two chain continuations are exactly opposite. For achiral polymers (R = R'), the sense of rotation of the dihedral angles is unimportant. But in chiral polymers, the physical situation for a torsional angle ϑ differs from the situation with a reversed sense, $-\vartheta$. In this review, we adopt the convention of Flory [12]. It is required that the direction of view is always from the substituted carbon atom to the methylene group CH_2. There is thus no constant direction of view. Then two situations may be distinguished, with the substituent R being either left of R' or *vice versa*. The sense of rotation of the dihedral angle changes accordingly (Fig. 10.b). This convention makes clear that for pseudoasymmetric polymers the bonds and not the atoms themselves are the generic origin of chirality; thus d- and l-bonds are discriminated. With this convention, the dihedral angles are unambiguously defined. Note, that irrespective of the tacticity, there are exactly as many d-bonds as there are l-bonds. Only the sequence of l's and d's depends on the configuration.

As noted above, the activation energy for constitutional and configurational transitions is much higher than thermal energies. Therefore, they do not regularly occur at room temperature. For conformations, however, the situation is different. Both energy differences and activation energies are of the order of 4 kJ/mol, thus comparable with thermal energies. Consequently, the conformational statistics is a dynamical and temperature dependent phenomenon with two characteristic features.

1. Assuming an energy curve like that in Fig. 2, only the regions around the relative energy minima $\vartheta = 0$, $\vartheta = \pm 120°$ will be appreciably populated. The actual dihedral angles found at any moment in a real sample

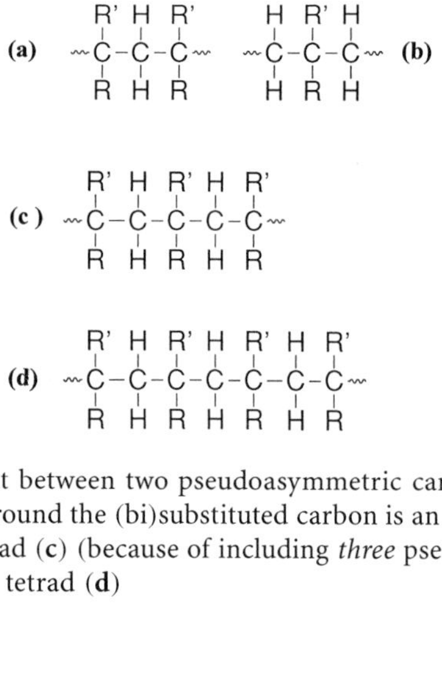

Fig. 9. A polymer segment between two pseudoasymmetric carbons is called a diad (**a**), and a segment centered around the (bi)substituted carbon is an interdiad (**b**). A sequence of two diads is called a triad (**c**) (because of including *three* pseudoasymmetric carbons), and a sequence of three a tetrad (**d**)

will scatter by a few degrees around these values. On the other hand, the probability of a high energy value $\vartheta = 180°$, $\vartheta = \pm 60°$ is negligible. The number, the positions and the energies of the minima depend on the specific polymer, its constitution and configuration, but the existence of a few, relatively well-defined regions of low energy (and high Boltzmann probability $\exp(-\beta E)$) holds for all systems. This finding is the basis for the Rotational Isomeric State (RIS) description of polymers [5], to be presented in the Sect. 2.2. The prominent values of the dihedral angles are frequently designated as *trans, t* ($\vartheta = 0°$), *syn, anti* ($\vartheta = \pm 60°$), *gauche, g* ($\vartheta = \pm 120°$), *cis, c* ($\vartheta = 180°$), though occasionally other terms are used.

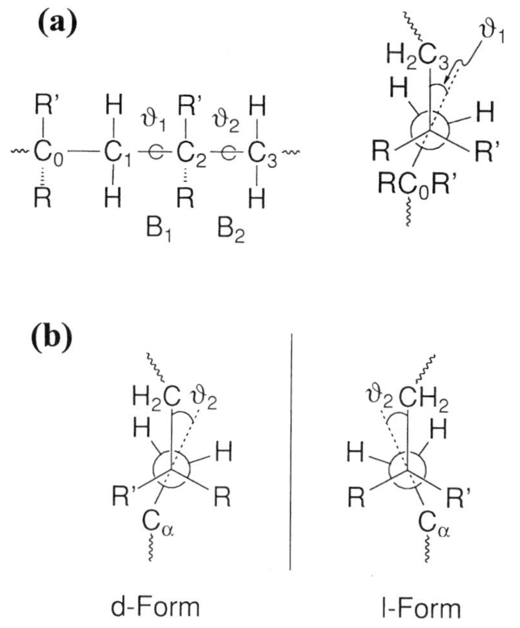

(a)

(b)

d-Form   l-Form

Fig. 10. The conformation of a molecule, i.e. the value of the dihedral angles ϑ is easily visualized by the *Newman*-projection. For the definition of the dihedral angle ϑ_1 (**a**), the line of view is from the bisubstituted carbon atom C_2 to the methylene carbon C_1. In contrast to ordinary organic chemistry, the dihedral angle ϑ_1 is assigned a value of $\vartheta_1 = 0°$ if the bonds C_2–C_3 and C_0–C_1 are exactly opposite.
If the two sidegroups R, R' of a bisubstituted vinyl polymer differ, the sense of rotation of the dihedral angles is important for the unambiguous assignment of a conformational state (**b**). After the convention of Flory [12], the direction of rotation depends on the configuration, as the line of view is from the pseudochiral carbon to the secondary carbon. Then, two situations may be distinguished, with R' being left of R, or *vice versa*

2. The energy barriers between different low-energy conformational states are low enough for thermally-induced conformational transitions. In solutions, dynamical equilibria are established with correlation times $\ll 1\ \mu s$ for conformational fluctuations. Consequently, solution NMR can measure only conformationally (but not configurationally) averaged spectra. Similarly, in the melt we find the same fast conformational transitions as in solution. But on cooling down of most polymers the viscosity increases tremendously, and the rate of conformational transitions decreases accordingly. Below the *glass transition*, the conformational disorder of the amorphous phase is frozen on the NMR time scale and NMR no longer records conformationally averaged but inhomogeneously broadened spectra due to the simultaneous presence of multiple conformations. In this review, we will be mainly concerned with the static local geometry of polymer glasses; for the still much debated dynamical aspects we refer to the literature [13, 3].

The overall shape of polymers in the amorphous phase corresponds to random coils. Their size, frequently expressed by the radius of gyration [14], can be inferred from neutron scattering experiments [15]. It has been shown, that (for a fixed polymer system) this global parameter does not vary much if measured in solution, in the melt or in the glassy state [16, 17, 18]. This demonstrates that the static properties of the molecular organization in the solution, the melt, and the glass are very similar at least on the length scale of the scattering experiments (> 10 Å). This is important for our analysis as the statistical models used for the simulation of the solid state NMR spectra are single chain models, taylored for ϑ-solvent conditions. By comparing simulations relying on such models with experimental solid state NMR spectra, we will be able to check whether the equivalence of molecular organization in the solution and the glass still holds on the strictly local scale of NMR.

2.1.1
Experimental Aspects

The experimental techniques for the recording of the 1D solid state spectra presented in the current work are standard [19, 3]. To get rid of the chemical shift anisotropy, Magic Angle Spinning (MAS) with a rotation frequency of at least 3000 Hz was applied throughout. For excitation either a Cross Polarization (CP) sequence or a single pulse was used. All spectra are ^{1}H-decoupled. They were either recorded on a BRUKER ASX-500 spectrometer (^{13}C -frequency 125 MHz) or a BRUKER MSL-300 (^{13}C -frequency 75.47 MHz). To exclude motional broadening or averaging, the recording temperature was adjusted to values below the respective glass transition temperature. Prior to the solid state NMR experiments, some of the polymers were characterized by ^{13}C solution NMR (BRUKER AC 300), by Differential Scanning Calorimetry (DSC) (METTLER), and by Wide Angle X-ray Scattering (WAXS).

2.2
Statistical Mechanics of Chain Molecules

In solution or in the amorphous solid state the geometry of a chain molecule cannot be exactly determined, but only be described by a set of *a priori* probability distribution functions. Important examples are the distribution of a single dihedral angle $p_1(\vartheta_1)$, the pair distribution function $p_2(\vartheta_1, \vartheta_2)$ and probability function for quadruples of dihedral angles $p_4(\vartheta_1, \vartheta_2, \vartheta_3, \vartheta_4)$.

We consider a chain of (N+1) carbon atoms (Fig. 11); sidegroups and hydrogen atoms have been omitted for the sake of simplicity. The set of coordinates for all atoms (including the hydrogens) is denoted by {**R**}. The conformation of the molecule is essentially determined by the dihedral angles $\vartheta_1, \ldots \vartheta_N$ (The chain end dihedral angles ϑ_1 and ϑ_N are not properly defined, and require a special treatment which is not addressed in this summary, as it is of merely technical interest). The energy of the chain molecule in Fig. 11 depends on the nuclear coordinates: $E(\{\mathbf{R}\})$. Then the probability for a single geometry {**R**} is (in principle) easily calculated by the laws of statistical mechanics.

$$p(\{\mathbf{R}\}) = \frac{1}{\mathscr{L}} \exp[-\beta E(\{\mathbf{R}\})], \qquad \beta = \frac{1}{k_B T} \qquad (2.1)$$

$$\mathscr{L} = \int d\{\mathbf{R}'\} \exp[-\beta E(\{\mathbf{R}'\})] \qquad (2.2)$$

Unfortunately, the calculation of the probability in the exact form Eq. (2.1) is a hopeless task.

A class of theories based mainly on the work of Flory [5] assumes certain simplifications of the energy hypersurface $E(\{\mathbf{R}\})$ which are introduced here in the most elementary form.

Fig. 11. Backbone of a polymer chain with $N + 1$ main chain carbon atoms; hydrogens and sidegroups have been omitted for simplicity. The conformation of the molecule is defined by the sequence of the dihedral angles $\vartheta_1, \ldots, \vartheta_N$. For the chain ends, a special treatment is required

1. Bond lengths and bond angles are kept constant.
2. Interactions between parts of the chain separated by more than 5 chain segments are neglected as long-distance excluded-volume interactions are effectively shielded for the mean chain.
3. The dihedral angles $\vartheta_1, \ldots, \vartheta_N$ adopt only a few discrete values $\Theta_1, \ldots \Theta_K$, the number K and the numerical values Θ_k depending on the specific polymer.
4. The energy $E(\{\mathbf{R}\})$ can be expressed as a function of the dihedral angles alone.

$$E\{(\mathbf{R})\} \longrightarrow E_{red}(\vartheta_1, \ldots, \vartheta_N) \tag{2.3}$$

$$= E_{red}(\Theta_{k_1}, \ldots, \Theta_{k_N}) \tag{2.4}$$

5. The energy function E_{red} can be decomposed into a sum of single dihedral and pair contributions.

$$E_{red}(\Theta_{k_1}, \ldots, \Theta_{k_N}) = \sum_{i=2}^{N} E_i^1(\Theta_{k_i}) + \sum_{i=2}^{N} E_i^2(\Theta_{k_{i-1}}, \Theta_{k_i}) \tag{2.5}$$

Because of the third assumption, these theories are called **Rotational Isomeric State (RIS) models**. Adopting these approximations, the partition function can be considerably simplified and the statistical mechanics of chain molecules can be mapped onto the well-known 1D Potts model with K states. [20]

$$\mathscr{L} \propto \sum_{\vartheta_1 = \Theta_1, \ldots \Theta_K} \cdots \sum_{\vartheta_N = \Theta_1, \ldots \Theta_K} \exp[-\beta E_{red}(\vartheta_1, \ldots \vartheta_N)] \tag{2.6}$$

$$= \sum_{k_1=1}^{K} \cdots \sum_{k_N=1}^{K} \exp[-\beta E_{red}(\Theta_{k_1}, \ldots, \Theta_{k_N})] \tag{2.7}$$

$$= \sum_{k_1=1}^{K} \cdots \sum_{k_N=1}^{K} \exp[-\beta \sum_{i=2}^{N} E_i^1(\Theta_{k_i}) + E_i^2(\Theta_{k_{i-1}}, \Theta_{k_i})] \tag{2.8}$$

$$= \sum_{k_1=1}^{K} \cdots \sum_{k_N=1}^{K} \prod_{i=2}^{N} \underbrace{\exp[-\beta(E_i^1(\Theta_{k_i}) + E_i^2(\Theta_{k_{i-1}}, \Theta_{k_i}))]}_{(U_i)_{k_{i-1}, k_i}} \tag{2.9}$$

$$= \sum_{k_1=1}^{N} \sum_{k_N=1}^{N} (U_2 \cdot U_3 \cdot \ldots \cdot U_N)_{k_1, k_N} \tag{2.10}$$

The partition function $\mathscr{L}$ is now simply a matrix product of weight matrices U_i; special care has to be devoted to the treatment of the endgroups.

The weight matrices U_i themselves can be inferred from several approaches with different levels of sophistication; in addition, it is sometimes necessary to relax the assumptions (1) to (5) [5, 21, 22, 23, 24, 25].

Mostly, the weight matrices are not calculated directly from the energies $E_i^{1,2}$ because the latter are often not known with the necessary accuracy. Instead, an inspection of the energy hypersurface leads to an effective parametrization of the weight matrices U_i, and the numerical values of the parameters are adjusted by comparison with experiment. Most of the statistical models in the literature are expressed in the form of suitable weight matrices together with a proper parametrization.

In a chain molecule, the same geometrical situation is repeated many times. Then, there are only a few different weight matrices. For example, for poly(ethylene) there is only one weight matrix, viz. $U_i = U \ \forall i$. In polymers with pseudoasymmetric centers, the weight matrices depend on the configuration, i.e. there are different matrices for m- and r-diads. By this mechanism, the configuration influences the conformational statistics. In terms of mathematics, the statistics is conditionally depending on the configuration. The observed configurational splitting in solution NMR spectra of polymers stems from this effect.

If the weight matrices are given, the probability distribution functions p_1, p_2, and p_4 are easily calculated in perfect analogy to the 1D-Ising model.

$$p_1(\vartheta_i = \Theta_l) \ = \ < \delta_{k_i,l} > \tag{2.11}$$

$$= \frac{1}{\mathscr{L}} \sum_{k_1=1}^{K} \cdots \sum_{k_i=1}^{K} \cdots \sum_{k_N=1}^{K}$$

$$(U_2)_{k_1 k_2} (U_3)_{k_2 k_3} \cdots (U_i)_{k_{i-1} k_i} \delta_{k_i l} (U_{i+1})_{k_i k_{i+1}} \cdots (U_N)_{k_{n-1} k_n}$$

$$= \frac{1}{\mathscr{L}} \sum_{k_1=1}^{K} \sum_{k_N=1}^{K} (U_2 \cdots U_{i-1} \hat{U}_i^{[l]} U_{i+1} \cdots U_n)_{k_1 k_n}$$

$\hat{U}_i^{[l]}$ is the matrix, obtained from U_i, if all columns but column "l" are set to zero. In the same way we get for the probability distribution of four consecutive dihedral angles

$$p_4(\vartheta_i = \Theta_{l_0}, \vartheta_{i+1} = \Theta_{l_1}, \vartheta_{i+2} = \Theta_{l_2}, \vartheta_{i+3} = \Theta_{l_3}) \tag{2.12}$$

$$= \ < \delta_{k_i,l_0} \cdot \delta_{k_{i+1},l_1} \cdot \delta_{k_{i+2},l_2} \cdot \delta_{k_{i+3},l_3} >$$

$$= \frac{1}{\mathscr{L}} \sum_{k_1=1}^{K} \sum_{k_N=1}^{K} (U_2 \cdots U_i \hat{U}_{i+1}^{[l_0,l_1]} \hat{U}_{i+2}^{[l_1,l_2]} \hat{U}_{i+3}^{[l_2,l_3]} U_{i+4} \cdots U_N)_{k_1 k_N} \ .$$

Here, $\hat{U}_i^{[l_0,l_1]}$ is the matrix U_i with all elements but (l_0, l_1) set to zero. The expression Eq. (2.12) is routinely used in the course of the simulation (*vide infra* Sects. 3, 5).

Some recent models avoid the calculation of weight matrices U_i and calculate the *a priori* pair probabilities $p^i(\Theta_{k_{i-1}}, \Theta_{k_i})$ directly [26, 27]. These approaches adopt an empirical force field energy $E(\{\mathbf{R}\})$ and apply a continuum Monte Carlo (MC) simulation [28] to obtain $p^i(\vartheta_{i-1}, \vartheta_i)$, $(\vartheta_{i-1}, \vartheta_i$ real). By integrating over suitable areas in the two-dimensional plane of dihedrals $(\vartheta_{i-1}, \vartheta_i)$, the results may be restated in terms of an RIS-like model and matrices of *a priori* probabilities are given:

$$p_2^i(\Theta_k, \Theta_l) = p_{kl}^i . \tag{2.13}$$

The index "i" refers to the geometrical situation of the bonds involved. There are as many different *a priori* matrices p_{kl}^i as there are different types of pairs of bonds. These calculations do not rely on special assumptions and poorly justified simplifications of the energy function; moreover, the approach is conceptually simple. The main disadvantage is the necessary computer time for the MC simulation, which has to be repeated for every temperature; at low temperature, the computer requirements become unaffordable.

If the *a priori* probabilities for pairs of consecutive dihedral angles are given, $p^i(\Theta_k, \Theta_l) = p_{kl}^i$, the probability distribution function for a sequence of n dihedral angles is easily calculated:

$$p_n(\Theta_{i_1}, \ldots, \Theta_{i_n}) = p_{i_1}^1 q_{i_1 i_2}^1 q_{i_2 i_3}^2 \cdots q_{i_{n-1} i_n}^n , \tag{2.14}$$

where

$$p_{i_1}^1 = \sum_{i_2} p^i(\Theta_{i_1}, \Theta_{i_2}) = \sum_{i_2} p_{i_1 i_2}^i \tag{2.15}$$

is the single bond a priori probability and

$$q_{i_1 i_2}^i = \frac{p_{i_1 i_2}^i}{\sum_k p_{i_1 k}^i} \tag{2.16}$$

is the *conditional* probability that $\vartheta_2 = \Theta_{i_2}$ is assumed if $\vartheta_1 = \Theta_{i_1}$ is given. Equation (2.14) tacitly assumes a Markovian character (only nearest-neighbor correlations) of the conformational statistics.

In the current work, we are not concerned with providing improved statistical models. Rather, we will use already existing models as given in the literature. By applying either Eq. (2.12) for a conventional RIS model or Eq. (2.14) for advanced Monte Carlo based models, we can infer the probability distribution $p_4(\vartheta_1, \ldots, \vartheta_4)$ which will be important for the simulation of the NMR spectra. On the other hand, global parameters (characteristic ratio, etc.) [5] are frequently calculated from RIS models for comparison with experiment. Because of this interplay of local and global geometry and to avoid a clumsy notation, "RIS model" and "random coil" picture are used quite indiscriminately in the current work.

2.3
Empirical Force Fields

The energy hypersurface $E(\{R\})$ of a molecule, i.e. the energy E as a function of the nuclear coordinates $\{R\}$, contains considerable information. The minima of the energy hypersurface represent the accessible stationary geometries, the curvature around these stationary points can be translated into infrared absorption spectra, its energy provides the Boltzmann factor, etc.

Unfortunately, the full energy hypersurface is not easily obtained. Even the determination of a single low energy stationary geometry by reliable ab initio quantum chemistry methods is very time consuming. For many applications it is thus necessary to replace the "exact" energy function $E(\{R\})$ by an effective parametrization, a so-called empirical force field $V_{eff}(\{R\})$ [29, 30, 31]. Here special effort is devoted to a correct representation of the low energy stationary states whereas the weakly populated high energy regions are less important. In this sense, an empirical force field is a kind of expansion of the energy hypersurface around the stationary points. Then the complicated energy function may be decomposed in a sum of easily interpreted terms.

- Contribution for stretching a bond by either a harmonic potential r_0 (*stretch*)

$$V_{stretch}(r) = \frac{1}{2}a(r - r_0)^2 \qquad (2.17)$$

  or a Morse function

$$V_{stretch}(r) = D_l[e^{-2\alpha(r-r_0)} - 2e^{-\alpha(r-r_0)}] . \qquad (2.18)$$

- Contribution for widening a bond angle α beyond its equilibrium value α_0

$$V_{bend}(\alpha) = \frac{1}{2}k(\alpha - \alpha_0)^2 \qquad (2.19)$$

- Contribution for the rotation around a bond

$$V_{torsion}(\vartheta) = \sum_i a_i[1 + cos(i(\vartheta - \vartheta_0))] \qquad (2.20)$$

  which is often simplified to a three-well-potential

$$V_{torsion}(\vartheta) = a_3[1 + cos(3(\vartheta - \vartheta_0))] \qquad (2.21)$$

- Coulomb interaction between atoms with partial charges q_i

$$V_{Coulomb}(r_{ij}) = \frac{1}{\epsilon}\frac{q_i q_j}{r_{ij}} \qquad (2.22)$$

ϵ is the dielectric constant, which may be distance-dependent.

- Van-der-Waals energy represented by a Lennard-Jones 6-12-potential

$$V_{vdW}(r_{ij}) = a[(\frac{f^*}{r_{ij}})^{12} - 2(\frac{f^*}{r_{ij}})^6] \tag{2.23}$$

- Terms for planarity in aromatic rings, hydrogen bonds, etc.

The total energy is then the sum of the various contributions.

$$V_{eff}(R_1, \ldots, R_K) = \sum_{bonds\ r_i} V^i_{stretch}(r_i) \tag{2.24}$$

$$+ \sum_{bond\ angles\ \alpha_k} V^k_{bend}(\alpha_k)$$

$$+ \sum_{dihedral\ angles\ \vartheta_l} V^l_{torsion}(\vartheta_l)$$

$$+ \sum_{i<j} V^{ij}_{Coulomb}(|\ R_i - R_j\ |)$$

$$+ \sum_{i<j} V^{ij}_{vdW}(|\ R_i - R_j\ |)$$

$$+ \ldots \tag{2.25}$$

The superscripts refer to different environments (and different parameters) within the molecule.

For the practical applicability, it is essential that the parameters used in the above equations depend primarily on the local bonding situation and may be transferred from one molecule to the other. Once the parameters are determined (by comparison with experiment or more sophisticated quantum chemistry calculations [32]), the energy function for a given molecule (or even group of molecules) may be obtained by a simple addition of the building blocks Eqs. (2.17)-(2.23) with the appropriate set of parameters.

In order to improve the representation of the energy hypersurface, modern empirical force fields contain so-called cross-terms that explicitly couple two hitherto independent parameters like two bond lengths.

$$V_{r_0, r_0'}(r, r') = F_{r_0, r_0'}(r - r_0)(r' - r_0') \tag{2.26}$$

If such terms are included and properly parametrized, fairly realistic geometries can be obtained.

In a well-parametrized force field a large number of different situations are distinguished. In the Consistent Valence Force Field (CVFF) of Biosym Inc. [33] used in the present paper, 15 different types of carbon atoms (like aromatic, carboxylic, etc.) are provided to permit a realistic energy function for a great variety of compounds. Prior to using an empirical force field it is absolutely essential to check whether an appropriate parametrization is available for the molecule in question. Subtle effects like partial conjugation are very often only poorly parametrized, and deserve special caution.

In this paper, the empirical force field will be used to generate equilibrium geometries, i.e. geometries $\{R_0\}$ with minimum energy

$$\nabla_{\{R\}} V_{eff}(\{R\}) \,|_{\{R_0\}} = 0 \tag{2.27}$$

The multiple minimum problem (of a high dimensional energy hypersurface) frequently encountered in large molecules is greatly mitigated by the fact that the dihedral angles will be fixed to their RIS values (*vide infra*, Sect. 3). Then standard minimization algorithms (steepest descent, conjugate gradient, Newton-Raphson) are sufficient for a proper optimization.

Most of the calculations to be presented here employ the CVFF empirical force field. It contains cross-terms and an extensive parametrization for the first two rows of the periodic system and some selected heavier atoms. It should provide reliable geometries for all polymer systems investigated. To assess the impact of the force field on the final result, the widespread AMBER forcefield [34, 35] was also used, which has a simpler parametrization and does not contain cross-terms. AMBER is mostly used for molecular dynamics simulations and should provide somewhat poorer geometries.

2.4
Ab Initio Quantum Chemical Calculation of the Chemical Shift by the IGLO Method

2.4.1
The Gauge Dependence of the Chemical Shift

In this section, we will briefly describe the ab initio computation of the chemical shift by the IGLO method of Kutzelnigg and Schindler [36, 37], which allows the investigation of the magnetic properties (chemical shift, magnetic susceptibility) of diamagnetic molecules. A detailed account with many applications has been published in this series [38]; technical aspects are covered in [39]. A new compact derivation in the language of quantum field theory is given in [40]; for some recent developments see also [41, 42]. A general introduction to quantum chemistry is given e.g. in [43, 44, 45]

In quantum mechanical terms, the chemical shift is the mixed second order perturbation of the energy of the molecule with respect to the exterior magnetic field B and the magnetic moment μ_k of the nucleus k in question

$$\sigma_{\alpha\beta}^k = \frac{\partial^2 E}{\partial \mu_\alpha^k \partial B_\beta}, \qquad \alpha, \beta = x, y, z \tag{2.28}$$

$E = \langle \mathcal{H} \rangle$ is the expectation value of the molecular Hamiltonian which consists of two parts, $\mathcal{H} = \mathcal{H}_0 + \mathcal{H}_1$, the ordinary Hamiltonian $\mathcal{H}_0$ and the magnetic perturbation $\mathcal{H}_1$.

We consider a molecule with N electrons and K nuclei. The ordinary electronic Hamiltonian of the unperturbed problem contains the kinetic energy of the electrons and the Coulomb interaction.

$$\mathcal{H}_0 = \sum_{n=1}^{N} \left(-\frac{\hbar^2}{2m_e} \nabla_{r_n}^2 - \sum_{k=1}^{K} \frac{Z_k e^2}{|\, r_n - R_k \,|} \right)$$
$$+ \sum_{m>n=1} \frac{e^2}{|\, r_n - r_m \,|} \tag{2.29}$$

In this formula, r_n denotes the cartesian coordinates of the nth electron, R_k the nuclear coordinates of nucleus k, and Z_k is the nuclear charge in units of the elementary charge e. The nuclear coordinates are given, and the Born-Oppenheimer separation has been performed.

The perturbation Hamiltonian $\mathcal{H}_1$ contains the magnetic perturbation which enter the Hamiltonian via minimal substitution of the linear momentum operator

$$\frac{\hbar}{\imath} \nabla_{r_i} \longrightarrow \frac{\hbar}{\imath} \nabla_{r_i} + \frac{e}{c} A(r_i), \tag{2.30}$$

where $A(r)$ is the vector potential associated with the total magnetic field

$$B(r) = \nabla_r \times A(r) \,. \tag{2.31}$$

The vector potential A is not defined unambiguously, but is subject to gauge transformations

$$A(r) \longrightarrow A'(r) = A'(r) + \frac{\hbar c}{e} \nabla_r \Lambda(r) \,, \tag{2.32}$$

where $\Lambda(r)$ is a smooth scalar function. Evidently, the results of a computation must not depend on the unphysical function Λ (gauge invariance).

For nonrelativistic calculations like the investigation of the chemical shift in light atom molecules, it is useful to impose the Coulomb gauge

$$\nabla_r \cdot A(r) = 0 \,. \tag{2.33}$$

But even then, a residual gauge freedom with scalar functions obeying $\nabla \cdot \nabla \Lambda = \Delta \Lambda = 0$ is preserved.

The magnetic field in Eq. (2.31) consists of the external static magnetic field B_0 and the dipole fields of the magnetic moments. Accordingly, the vector potential is the sum of these contributions.

$$A(r) = \frac{1}{2} B_0 \times (r - R) + \sum_{k=1}^{K} \frac{\mu_k \times (r - R_k)}{|\, r - R_k \,|^3} \tag{2.34}$$

Here, R is an arbitrary *gauge origin* that can be changed by a special gauge transformation

$$\Lambda_{RR'}(r) = \frac{e}{2\hbar c}(B_0 \times (R - R')) \cdot r \tag{2.35}$$

$$A'(r) = A(r) + \nabla_r \Lambda_{RR'}(r)$$

$$= \frac{1}{2}B_0 \times (r - R') + \sum_{k=1}^{K} \frac{\mu_k \times (r - R_k)}{|r - R_k|^3} \tag{2.36}$$

In a molecule we are free to select any value R, since there is no obvious choice for it (as it is the nuclear position in an atom). The free gauge origin R is the principal expression of gauge freedom in the quantum mechanical theory of the chemical shift.

Inserting the Eqs. (2.30), (2.34) into Eq. (2.29) provides the perturbation Hamiltonian $\mathcal{H}_1$ which is a very complicated function that contains terms linear in B_0, linear in μ_k, bilinear in B_0 and μ_k, and bilinear in μ_k and $\mu_{k'}$.

The computation of the chemical shift (2.28) is a formidable task. In a first step, $\mathcal{H}_1$ is neglected, and a Hartree-Fock calculation for the unperturbed problem $\mathcal{H}_0$ is performed. The solution of the Hartree-Fock calculation is a single Slater determinant[1]

$$\Psi(q_1, \ldots, q_N) = \frac{1}{\sqrt{N!}} \begin{vmatrix} \chi_1(q_1) & \cdots & \chi_N(q_1) \\ \vdots & \ddots & \vdots \\ \chi_1(q_N) & \cdots & \chi_N(q_N) \end{vmatrix} \tag{2.37}$$

$$= \frac{1}{\sqrt{N!}} \det(\chi_1 \cdots \chi_N), \tag{2.38}$$

that minimizes the energy expectation value

$$E[\Psi] = \langle \Psi \mid \mathcal{H}_0 \mid \Psi \rangle \tag{2.39}$$

within the set of accessible Slater determinants:

$$\frac{\delta E[\Psi]}{\delta \Psi} = 0. \tag{2.40}$$

In practical calculations, the canonical Molecular Orbitals (MO) χ_i are expressed as a sum of Atomic Orbitals (AO) ϕ_k^j centered around the nucleus of the atoms.

$$\chi_i(q_i) = \sum_{k=1}^{K} \sum_{j=1}^{n_k} a_{ik}^j \phi_k^j(q) \tag{2.41}$$

---

[1] For the time being, q_i is a unified notation for both spatial (r_i) and spin (s_i) coordinates, $q_i = (r_i, s_i)$.

The *ansatz* Eq. (2.41) is the Linear Combination of Atomic Orbitals (LCAO) approach. The coefficients a_{ik}^{j} are the parameters of variation when minimizing the functional $E[\Psi]$. The set of functions belonging to a single atom $\{\phi_k^{j} \mid j = 1, \ldots, n_k\}$ is called atomic basis set.

A Hartree-Fock solution is only approximate, Ψ_{HF} does *not* solve the stationary Schrödinger equation. Thus,

$$\mathcal{H}_0 \Psi_{HF} \neq E_{HF} \Psi_{HF} \tag{2.42}$$

as the correlation of electrons (to be encoded by a sum of Slater determinants) has been neglected (*correlation* effect). The reliability of the Hartree-Fock solution depends crucially on the atomic basis set (*vide infra*).

Having approximately solved the unperturbed problem on the Hartree-Fock level, a second-order perturbation calculation in $\mathcal{H}_1$ is necessary to obtain the chemical shift. A straightforward computation is faced with two severe problems:

1. A perturbation calculation starting from an approximate wave function (such as Ψ_{HF}) is susceptible to ambiguities and large errors (see [39, p. 18]). In order to obtain reliable results, it is necessary to stabilize the solutions with respect to a certain class of variations. This approach is called *stationary perturbation theory*; its Hartree-Fock level version is called Coupled Hartree-Fock (CHF).
2. In an approximate calculation, it is difficult to preserve gauge invariance (which is naturally fulfilled in an exact approach). Thus, the result will depend on the gauge origin R, which bears no physical relevance. Gauge-dependent chemical shifts cannot be relied upon, as they violate fundamental physical principles.

As a consequence of these two problems, a naive calculation will yield chemical shifts which are small differences of large numbers, and which are consequently subject to large numerical uncertainties. These problems can be remedied by the IGLO approach. In this method, the canonical molecular orbitals χ_i are transformed into localized molecular orbitals φ_j by means of a unitary matrix U

$$\varphi_i = \sum_n U_{in} \chi_n \tag{2.43}$$

The localization procedure follows the prescription of Boys [46, 47, 48, 49] and leads to Localized Molecular Orbitals (LMO) φ_i that bear physical meaning. These orbitals are centered around nuclei or bonds and are thus easily interpreted in the way of textbook chemistry. The essence of the IGLO method is the choice of an Individual Gauge for (each) Localized Molecular Orbital, which explains the abbreviation. The gauge origin R is set to the centre of charge of every LMO independently. Additionally, the stationarity of the solution is assured by CHF conditions. The IGLO scheme avoids the problems related with a straightforward naive calculation and has been successfully applied to a large number of compounds [38]. It has been very useful in the elucidation of the structure of samples for which increment systems were not available. Moreover, the IGLO method permits the identification of MO contributions to the chemical shift and the calculation of the full chemical shift tensor.

In the past few years, two other methods have been increasingly used to obtain the chemical shift on the Hartree-Fock level.

1. Gauge Including Atomic Orbitals[2] uses local gauge origins as well, but different from IGLO these are applied to atomic orbitals rather than localized molecular orbitals. The principal idea of GIAO dates back to London [50]. Later, Pople [51, 52], Hameka [53], and Ditchfield [54] applied GIAO to quite small molecules. Only a new efficient implementation by Wolinski, Pulay and Hinton [55] allowed the treatment of larger molecules. Some new results are given in [56].

2. Local Origin/Local Gauge LORG has been developed by Hansen and Bouman [57], and like IGLO uses gauge dependent localized molecular orbitals. It differs from IGLO in a subtle way by violating a stationarity condition [39]. Some new results are presented in [58].

The three concepts (IGLO, GIAO, LORG) have been first implemented on the Hartree-Fock level; effects of electron correlation have been neglected. In the past few years, beyond-Hartree-Fock extensions have been developed: a Multi-Configuration Self-Consistent Field (MC-SCF) version of IGLO [40, 42], a perturbation treatment of correlation effects for GIAO [59, 60] and LORG [58]. These extensions are particularly important for calculations of absolute chemical shifts in compounds with presumably strong electron correlation such as molecules with π-bonds or heavy atoms.

The computations of conformational effects on the chemical shift in this study have been performed with the semi-direct version of ordinary Hartree-Fock IGLO [41].

2.4.2
Basis Sets

As already mentioned, in the LCAO approach the chosen atomic basis is essential for the validity of the results. For IGLO, some basis sets have become standard. All are Huzinaga sets [61] of Gaussian functions. The notation of [38] is used.

---

[2]Originally, the abbreviation read "Gauge *invariant* atomic orbitals" which is very confusing and should be no longer used.

Table 2. Characteristics for the *double zeta* basis DZ employed in the IGLO calculations. DZ is a Huzinaga basis (second column) with a special contraction scheme (third column). The resulting number of functions (fifth column) depends on the contracted basis (fourth column). It should be noted, that "*p*-function" is shorthand for a set of three functions (p_x, p_y, p_z), and "*d*-function" for a set of five, etc. [45, 43, 44]

DZ	(s, p)	contraction	contracted basis	number of functions
H	$(3s, 0p)$	$(2,1;0)$	$[2s, 0p]$	$2 \cdot 1 = 1$
C	$(7s, 3p)$	$(4,1,1,1;2,1)$	$[4s, 2p]$	$4 \cdot 1 + 2 \cdot 3 = 10$
O	$(7s, 3p)$	$(4,1,1,1;2,1)$	$[4s, 2p]$	$4 \cdot 1 + 2 \cdot 3 = 10$
Cl	$(10s, 6p)$	$(5,1,1,1,1,1;3,1,1,1)$	$[6s, 4p]$	$6 \cdot 1 + 4 \cdot 3 = 18$

Table 3. Characteristics for basis II, which is of *triple zeta* quality. It is improved by so-called polarization functions, which are *p*- and *d*-type functions, respectively

II	(s, p, d)	contraction	contr. basis	polarization functions	number of functions
H	$(5s, 1p, 0d)$	$(3, 1, 1; 1; 0)$	$[3s, 1p, 0d]$	$1p, \xi = 0.65$	$3 \cdot 1 + 1 \cdot 3 = 6$
C	$(9s, 5p, 1d)$	$(5, 4 \times 1; 2, 3 \times 1; 1)$	$[5s, 4p, 1d]$	$1d, \xi = 1.0$	$5 \cdot 1 + 4 \cdot 3 + 1 \cdot 5 = 22$
O	$(9s, 5p, 1d)$	$(5, 4 \times 1; 2, 3 \times 1; 1)$	$[5s, 4p, 1d]$	$1d, \xi = 1.0$	$5 \cdot 1 + 4 \cdot 3 + 1 \cdot 5 = 22$
Cl	$(11s, 7p, 2d)$	$(5, 6 \times 1; 2, 5 \times 1; 1, 1)$	$[7s, 6p, 2d]$	$2d, \xi = 0.4, 1.6$	$7 \cdot 1 + 6 \cdot 3 + 2 \cdot 5 = 35$

- DZ denotes a basis of "*double-zeta*"-quality, i.e. for each electron of the shells involved there are two basis functions. In Table 2 the basis sets and contraction schemes are listed for the elements H, C, O, Cl which will suffice for the investigated systems.
- II is a basis of *triple-zeta* quality (i.e. three functions per electron) enhanced by polarization functions. Table 3 shows the characteristics of this basis.
- II' denotes a basis where the heavy atoms (C, O, Cl) have been equipped with II and the hydrogen atoms with DZ.
- Other basis sets (denoted by I, III, and IV in [38]) are either obsolete or too large for the available computer resources (*vide infra*).

On selecting an atomic basis for a calculation, two considerations must be balanced.

1. The necessary computing power increases with the 3rd to 4th power of the overall number of atomic basis functions. Doubling the atomic basis (e.g. by switching from DZ to II') roughly increases the computing time by one order of magnitude.
2. As a rule of thumb, the results are the more reliable the larger the basis.

For the analysis and prediction of solution spectra, normally at least the basis II is required. DZ can be used for hydrocarbons [38, Chap. 2.15]. However, in this study, we are not interested in the absolute position of the resonances but in the geometry-induced variations, i.e. relative effects. Then many errors of a small atomic basis cancel and DZ provides reliable results for most of the systems discussed (*vide infra*).

2.4.3
Comparison of Experiment and Theory

As can be seen from Eq. (2.28), the chemical shift is a general 2nd rank tensor. In NMR spectra, only the symmetric part

$$\bar{\sigma}_{\alpha\beta} = \frac{1}{2}(\sigma_{\alpha\beta} + \sigma_{\beta\alpha}) \tag{2.44}$$

is directly observable. For simplicity, in the following, the symmetrized tensor $\bar{\sigma}_{\alpha\beta}$ will be denoted by the same symbol $\sigma_{\alpha\beta}$. For liquid spectroscopy and CP/MAS solid state spectra, only the isotropic part

$$\sigma := \sigma_{iso} = \frac{1}{3}(\sigma_{xx} + \sigma_{yy} + \sigma_{zz}) \tag{2.45}$$

is of interest. The anisotropy of the chemical shift can be investigated by special experiments (Sect. 3.8); then the principal tensor components σ_{xx}^{PT}, σ_{yy}^{PT}, and σ_{zz}^{PT} can be inferred.

When comparing experimental and theoretical data, it is essential to note that theoretical calculations yield absolute values with respect to the bare nucleus *in vacuo* whereas experimental data refer to a reference (Tetra-Methyl-Silane (TMS) for ^{13}C). The experimental scale δ is related to the theoretical scale σ via

$$\delta = \sigma_0 - \sigma \tag{2.46}$$

where σ_0 is the theoretical value of the reference. In practice, the shortcomings of a still approximate quantum chemical calculation and the use of small atomic bases leads to errors in the absolute δ-values of the various resonances (For problems of principal character see [38, Sect. 3.1]). But in this study, we focus on relative effects, i.e. the relative shift of the resonances of the same molecule due to its change in geometry, or the configurational splitting. For these questions, the error in the absolute values are not very essential (*vide infra*).

2.5
Geometry Effects in NMR Spectra – the γ-gauche Effect

The most important geometrical effect on NMR spectra of chain molecules is the γ-gauche effect and has already been discussed in the Introduction. The experimental effect has been parametrized by a phenomenological theory

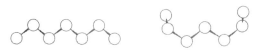

Fig. 12. The chemical shift of the central carbon atom C_0 depends on the position of its γ-neighbors X_1 and X_2. If the dihedral angles ϑ_1 and ϑ_2 switch from *trans* to *gauche*, the carbon atom C_0 is more shielded, and the chemical shift δ is reduced

that is now shortly reviewed. If the dihedral angle ϑ_1 in Fig. 12 changes from *trans* to *gauche*, the distance $C_0 - X_1$ is reduced by ≈ 1 Å, the carbon C_0 is accordingly more shielded, and its chemical shift is reduced by $\gamma_{X_1} \approx 5$ ppm. If the second γ-neighbor X_2 is in a *gauche* position, too, the shielding is additive, and the chemical shift of C_0 differs by $\gamma_{X_1} + \gamma_{X_2}$ from the stretched geometry with $\vartheta_1 = t$, $\vartheta_2 = t$.

$$\delta_{C_0}(\vartheta_1 = g, \vartheta_2 = t) = \delta_{C_0}(\vartheta_1 = t, \vartheta_2 = t) + \gamma_{X_1}, \quad \gamma_{X_1} < 0 \qquad (2.47)$$
$$\delta_{C_0}(\vartheta_1 = g, \vartheta_2 = g) = \delta_{C_0}(\vartheta_1 = t, \vartheta_2 = t) + \gamma_{X_1} + \gamma_{X_2}, \quad \gamma_{X_i} < 0 \ (2.48)$$

The γ-*gauche* constants γ_{X_i} are empirical parameters and have to be adjusted by comparison with experiment.

The above formulae tacitly assume that the carbon atoms are exactly tetrahedrally coordinated and that the γ-*gauche* effect is of exclusively steric origin. Other geometrical parameters, especially dihedral angles other than ϑ_1 and ϑ_2, are not expected to have an impact on the chemical shift. This assertion is at variance with the observed experimental solid state spectra, that display broad resonances rather than sharp peaks (see Introduction). In Sects. 3 and 5 it is shown by quantum chemical calculations, that Eqs. (2.47), (2.48) hold only approximately and have to be modified considerably for real systems. In fact, for poly(methyl methacrylate) (PMMA) a complete breakdown of the γ-*gauche* effect will be reported.

In spite of the shortcomings of the phenomenological Eqs. (2.47), (2.48), they can be used for a successful explanation of the configurational splitting in solution spectra [62, 4, 8]. As noted above, in solution, fast conformational transitions occur and only conformationally averaged resonance lines can be detected. If the probability of finding X_1 in a *gauche* position is given by $P_g^{X_1}$ (which may be calculated from a suitable RIS model), the corresponding resonance line will be shifted by

$$\Delta\delta = \gamma_{X_1} P_g^{X_1} \qquad (2.49)$$

with respect to the *trans* position. If there are multiple γ-neighbors, their influence is additive

$$\Delta\delta = \sum_i \gamma_{X_i} P_g^{X_i} \ . \tag{2.50}$$

The conformational splitting occurs because the probability $P_g^{X_i}$ depends on the configuration; the *γ-gauche* constant itself is assumed to be insensitive to configuration. The probability is sensitive to larger chain segments than the direct quantum mechanical *γ-gauche* effect itself, as relatively long configurational sequences influence the probability $P_g^{X_i}$; splittings due to different hexads are frequently observed. In Fig. 13 (reprinted for reader convenience from [63]), solid state and solution NMR spectra of atactic poly(propylene) are compared. The solution spectrum (bottom) of the methylene carbons spreads over a region of less than 3 ppm. The solid state resonance is much broader (top) and spreads over a region of up to 20 ppm. The reason is obvious: Solid state spectra display (in an indirect manner) a *distribution* of geometries, solution spectra *average* over the conformational distributions and only the relatively small differences in the mean probability $P_g^{X_i}$ of finding X_i in a *gauche* state show up.

In the past, the *γ-gauche* effect has been mostly used as an empirical concept for spectral assignment (see e.g. [64]). Attempts to understand the reasons behind the phenomenon have been rare [65]. Only in the last few years, quantum chemical methods have been used for a quantitative investigation of conformational effects mainly in small organic molecules and biomolecules [66, 67, 68, 69, 70, 71, 72, 73, 74]. To our knowledge, no investigation had been devoted to conformational effects in synthetic polymers previous to our calculations [9, 11, 10]. However, more recently, a paper investigating models of poly(vinyl alcohol) by GIAO [75] has been published.

The ^{13}C chemical shift is not the only NMR parameter that displays a geometry dependence. Comparable effects have been reported for ^{15}N and ^{19}F [72]. Proton shifts are typically too small for an analysis [8]. The vicinal coupling $^3J_{H,H}$ depends heavily on the dihedral angles; this dependence has been parametrized by the Karplus curve. It is widely used for the conformational and configurational analysis of low molecular compounds, especially for ring systems with restricted conformational mobility. Moreover, J-couplings are used as a source of structural information [76]. In solid state NMR, the vicinal coupling and its geometry dependence is normally unimportant.

In oxidic mineral glasses, a number of investigations [77, 70, 78] have shown that ^{31}P and ^{29}Si chemical shifts depend in a systematic fashion on bond lengths and bond angles – geometry parameters that are essential for the understanding of the amorphous disorder in these systems (*vide infra* Sect. 4.6).

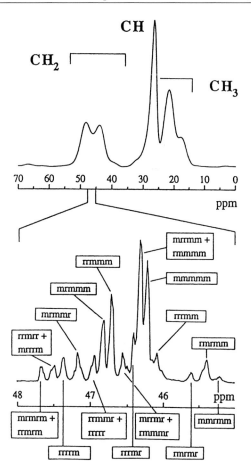

Fig. 13. Experimental solid state spectrum of atactic poly(propylene) (*top*) and solution spectrum of configurational splitting in the methylene region (*bottom*). The broadening in the solid state spectrum is appreciably larger than the configurational splitting. Reprinted with minor changes from [63]

In principle, all NMR parameters depend on geometry. For chemical shifts and indirect couplings, the influence is rather well-behaved and can be used for structural investigations. In principle, the geometry can directly be probed by the dipolar coupling between $^1H - ^1H$ and $^1H - ^{13}C$ pairs. However, the multibody nature of the dipolar coupling in solids has so far precluded this approach. The advent of high resolution multiple quantum solid state NMR spectroscopy [79, 80, 81, 82] should make it possible to use the dipolar coupling as a source of structural information and probe the local geometry in a direct way without extensive isotopic labelling [83, 84].

3
Method of ab initio Simulation of Solid State NMR Spectra –
The Example of Atactic Poly(propylene)

3.1
Atactic Poly(propylene)

In this section, the method underlying the ab initio simulation of chemical shifts of amorphous polymers will be explained in some detail. In order to avoid a clumsy general notation, we will refer to the specific example of atactic poly(propylene) (aPP). The aPP sample used as experimental reference has an almost perfect Bernoullian configurational statistics ($p(m) = p(r) = 0.5$, $p(mm) = p(rr) = 0.25$, $p(mr) + p(rm) = 0.5$, etc. are fulfilled within experimental errors [85, 63]). The CP/MAS spectrum of the polymer glass and the solution spectrum have been discussed in the previous section (Fig. 13). Conformational broadening and configurational splitting are clearly visible. In the solid state spectrum, the methylene resonance is richly structured and displays two main contributions and a high field shoulder. The methyl peak is structured, too, whereas the methine carbon is comparatively narrow. The high spinning speed (3.4 kHz) precludes any broadening due to residual chemical shift anisotropy. Motional broadening can be excluded by 2D exchange spectroscopy [63]. An experiment with a long mixing time (0.5 s) does not show magnetization exchange; consequently conformational transitions are absent on this timescale.

3.2
Statistical Model

For the simulation both of the solid state NMR spectrum and the configurational splitting [10] in solution we adopt the 5-state RIS model of Suter and Flory [86]. This model correctly reproduces the experimental radii of gyration; some difficulties have been reported for various temperature coefficients [87]. For amorphous isotactic poly(propylene) the radii of gyration in solution and in the melt nearly coincide [88, 18]. Thus, the model [86] is apparently a valid description of poly(propylene) both in solution and in bulk on the longer length scale (> 10 Å) of scattering experiments.

The local conformational statistics has not been probed directly by an experiment, but a molecular mechanics simulation by Theodorou and Suter

Table 4. Rotational isomeric states for the statistical model [86] of poly(propylene). The sense of rotation of a dihedral angle depends on the chirality of the atoms involved

Name	t	t^*	g^*	g	$\bar{g}$
ϑ	$15°$	$50°$	$70°$	$105°$	$-115°$

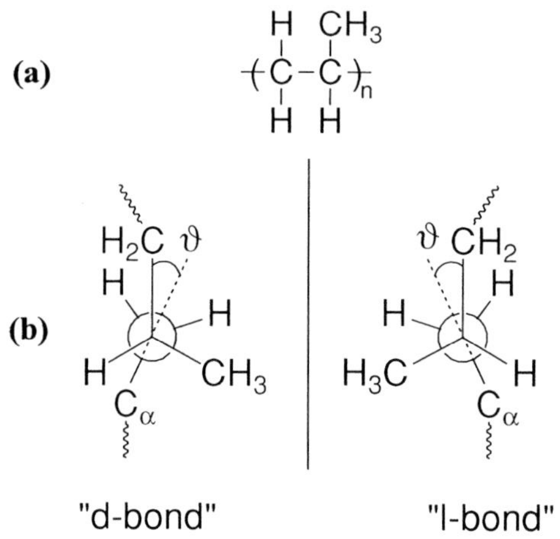

Fig. 14. Structural formula of poly(propylene) (**a**), which has pseudoasymmetric carbons. Definition of dihedral angles adapted to the case of poly(propylene) (**b**) (cf. Fig. 10.b)

[89] shows a semiquantitative agreement with the RIS model. Some minor discrepancies may be either due to errors in the RIS model or due to idealizations in the molecular mechanics simulations. Considering all findings, the RIS model seems to be a reasonable description of poly(propylene) even on the local length scale. It is thus a good candidate for the testing of our approach as we can be fairly sure about the statistical model. A more detailed discussion of the role of the conformational statistics will be presented in Sect. 4.4.

The model assumes five states listed in Table 4. The torsional angles are measured according to the convention described in Sect. 2.1; the sense of rotation of the dihedral angle depends on the configuration (Fig. 14.b).

The overall geometry of the macromolecule depicted in Fig. 15.a is determined by the sequence of the diads (*mrmmr* in the figure) and the sequence of dihedral angles

$$\cdots \mid \vartheta_{i-1} \vartheta_i \mid \vartheta_{i+1} \vartheta_{i+2} \mid \vartheta_{i+3} \vartheta_{i+4} \mid \cdots \qquad (3.1)$$

where the vertical bar (|) denotes the position of the substituted chiral methine carbons. The weight matrices depend on the configuration. For

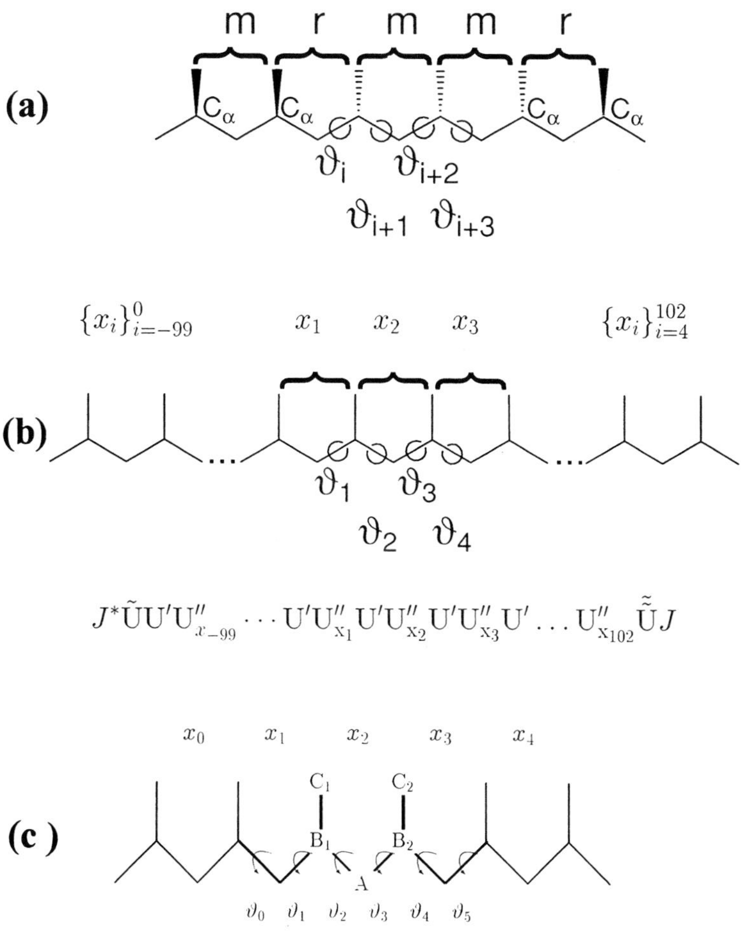

(a)

(b)

(c)

Fig. 15. Segment of an extended poly(propylene) chain. The overall geometry of the molecule depends both on the configuration (*mrmmr*) and the conformation (ϑ_i) (**a**). The relevant conformational statistics is determined by considering the central part of an extended polymer chain (**b**). The statistical weight of a chain molecule with a fixed configuration $\{x_i\}$ is determined by multiplying the appropriate RIS weight matrices. The weight matrices depend on the configuration.

The actual quantum chemical calculations rely on a model molecule (**c**), a hexamer with methyl endgroups and many degrees of freedom. The γ-neighbors of the central methylene (A) and methyl (C_1, C_2) units are fixed by defining the central diads x_1, x_2, x_3 and the dihedral angles $\vartheta_1, \vartheta_2, \vartheta_3, \vartheta_4$. These parameters have to be set to fixed values; then a region of well-defined geometry within the model molecule is established; it is depicted by bold lines. For the γ-neighbors of the methine units (B_1, B_2) the additional dihedral angles ϑ_0 and ϑ_5 should be set to defined values, too. As the experimental resonance of the methine unit (Fig. 5) is very narrow and of little interest, this computer-time consuming refinement has been omitted

meso-diads the weight matrix reads

$$
U''_m = \begin{pmatrix}
0 & \eta\eta^*\omega & 0 & \eta & 0 \\
\eta\eta^*\omega & 0 & 0 & 0 & \eta^*\tau\omega \\
0 & 0 & 0 & \eta^*\omega & \eta^*\tau\omega \\
\eta & 0 & \eta^*\omega & 0 & 0 \\
0 & \eta^*\tau\omega & \eta^*\tau\omega & 0 & 0
\end{pmatrix}
\tag{3.2}
$$

und for a *racemic* diad

$$
U''_r = \begin{pmatrix}
\eta^2 & 0 & \eta\eta^*\omega & 0 & 0 \\
0 & 0 & 0 & \eta^*\omega & \eta^*\tau\omega \\
\eta\eta^*\omega & 0 & 0 & 0 & \eta^*\tau\omega \\
0 & \eta^*\omega & 0 & 1 & 0 \\
0 & \eta^*\tau\omega & \eta^*\tau\omega & 0 & 0
\end{pmatrix}.
\tag{3.3}
$$

These matrices show explicitly how the configuration acts on the conformation. By appointing different configuration-depending weights to the various
states, the probability of finding a bond in *gauche* (or any other) state is af-
fected, and a different, configuration-dependent conformational statistics is
observed. The weight matrix of an interdiad (like $\vartheta_i|\vartheta_{i+1}$) does not show a
configurational dependence.

$$
U' = \begin{pmatrix}
1 & 1 & 1 & 1 & 1 \\
1 & 1 & 1 & 1 & 1 \\
1 & 1 & 0 & 0 & 1 \\
1 & 1 & 0 & 0 & 1 \\
1 & 1 & 1 & 1 & 0
\end{pmatrix}
\tag{3.4}
$$

The parameters of the weight matrices are given in [86],

$$
\eta^*\omega = 0.9 \exp(-6698 \text{ J mol}^{-1}/RT)
\tag{3.5}
$$
$$
\tau = 0.4 \exp(-2093 \text{ cal mol}^{-1}/RT)
\tag{3.6}
$$
$$
\eta = 1.0 \exp(-251 \text{ cal mol}^{-1}/RT)
\tag{3.7}
$$

By applying Eq. (2.10) for the evaluation of the partition function and
Eq. (2.12) for the computation of the conformational probabilities, these matrices describe the local and global geometry of the random coil polymer
chain.

3.3
Simulation of the Spectra

3.3.1
Identification of Important Conformations

The model shortly reviewed in the previous section will be the base for the simulation of the NMR spectrum. A sequence of three specified diads x_i and four well-defined dihedral angles ϑ_i will be necessary for the simulation (*vide infra*). As we want to investigate a high molar mass polymer and not a low molecular compound, we concentrate on a small sequence within an extended polymer chain of several hundred monomeric units (Fig. 15.b). We are interested in the probability distribution of the four dihedral angles ϑ_1, ϑ_2, ϑ_3, ϑ_4, if the configuration x_1, x_2, x_3 of the three central diads is given and the rest of the chain has the experimental configurational statistics (which is ideally atactic in this case). The expression looked for is thus the conditional probability

$$p(\vartheta_1, \vartheta_2, \vartheta_3, \vartheta_4 \mid x_1, x_2, x_3) \qquad (3.8)$$

The dihedral angles ϑ_i can be chosen out of the five rotational isomeric states $\{t, t^*, g^*, g, \bar{g}\}$, and the diads are either *meso* or *racemic*, $x_i \in \{m, r\}$. For the simulation, this probability distribution is needed for all configurational sequences $(x_1, x_2, x_3) = (m, m, m), (m, m, r), \ldots, (r, r, r)$.

The configuration of the whole extended chain is given by the sequence

$$x_{-99} \cdots x_0 x_1 x_2 x_3 x_4 \cdots x_{102}, \qquad x_i \in \{m, r\} \qquad (3.9)$$

The central diads (x_1, x_2, x_3) have fixed values whereas the others are determined by a random number. The probabilities for the selection of an m- or r-diad are set to the ideally atactic case $p_m = p_r = 0.5$. Then the partition function for such a chain is

$$\mathcal{Z}(\{x_{-99} \cdots x_1 x_2 x_3 \cdots x_{102}\}) = J^* \tilde{U} \left[\prod_{k=-99}^{102} U' U''_{x_k} \right] \tilde{U} J . \qquad (3.10)$$

The matrices $J^*, J, \tilde{U}$, and $\tilde{U}$ are special matrices for the correct representation of the chain ends [86, Eq. (17)-(20)]. The conditional probability distribution for a chain with a given configurational sequence $\{x_{-99} \cdots x_{102}\}$ is calculated by a specialized version of Eq. (2.12).

$$p_4(\vartheta_1, \vartheta_2, \vartheta_3, \vartheta_4 \mid x_{-99} \cdots x_1 x_2 x_3 \cdots x_{102}) \qquad (3.11)$$

$$= \frac{1}{\mathcal{Z}(\{x_i\})} J^* \tilde{U} \left(\prod_{k=-99}^{1} U' U''_{x_k} \right) \left[\hat{U}'^{[\vartheta_1, \vartheta_2]} \hat{U}''^{[\vartheta_2, \vartheta_3]}_{x_2} \hat{U}'^{[\vartheta_3, \vartheta_4]} \right] U''_{x_3} \left(\prod_{k=4}^{102} U' U''_{x_k} \right) \tilde{U} J$$

In the matrices $\hat{U}^{[\vartheta,\vartheta']}$ all elements but (ϑ,ϑ') are set to zero.

The probability $p(\vartheta_1, \vartheta_2, \vartheta_3, \vartheta_4 \mid x_1, x_2, x_3)$ is obtained by taking the average over many (typically $N=10000$) chains with a random configuration $\{x_i^k\}$ where only (x_1^k, x_2^k, x_3^k) are set to fixed values (x_1, x_2, x_3).

$$p_4(\vartheta_1, \vartheta_2, \vartheta_3, \vartheta_4 \mid x_1, x_2, x_3) = \frac{1}{N} \sum_{k=1}^{N} p_4(\vartheta_1, \vartheta_2, \vartheta_3, \vartheta_4 \mid |x_{99}^k...x_1, x_2, x_3...x_{102}^k)$$

(3.12)

The convergence $N \to \infty$ was checked by varying N; for all systems investigated $N \approx 10000$ was by far sufficient. In the same way, it has been established that the use of longer chains with more (1000, 2000) monomeric units does not affect the results. As far as conformational statistics is regarded, the central part of a chain with 200 monomeric units does not deviate from the central part of a larger chain molecule. Endgroup effects are negligible. The probability distribution Eq. (3.12) fulfills an important symmetry relation

$$p_4(\vartheta_1, \vartheta_2, \vartheta_3, \vartheta_4 \mid x_1, x_2, x_3) = p_4(\vartheta_4, \vartheta_3, \vartheta_2, \vartheta_1 \mid x_3, x_2, x_1),$$

(3.13)

which reflects the irrelevance of the direction from which the chain is seen.

An actual calculation shows that out of $2^3 \cdot 5^4 = 5000$ possible sequences $k^j = (\vartheta_1^j, \vartheta_2^j, \vartheta_3^j, \vartheta_4^j; x_1^j, x_2^j, x_3^j)$ only about 140 contribute with an appreciable probability > 0.001; the overall probability of these 140 sequences is > 0.91 at $T = 260$ K. This is a great asset for the ab initio spectrum simulation as only relatively few conformations must be taken into account explicitly whereas the by far larger number can be safely omitted.

3.3.2
Geometry Optimization and Quantum Chemistry

The relevant conformations have now been determined and for each of these conformations the chemical shifts are required. Quantum chemical calculations of an extended polymer chain like in Fig. 15.b are impossible in practice because of the computer time needed. Thus we have to find a small model molecule for quantum chemistry calculations (Fig. 15.c). For the selection of this model molecule, two contradicting requirements must be balanced.

1. The necessary computer time increases with the 3rd to the 4th power of the size of the atomic basis, which is approximately proportional to the number of non-hydrogen nuclei. For a fast calculation, a small molecule is needed.
2. As a simulation of a macromolecule is intended, the central part of the model molecule (denoted by the atoms A, B_1, B_2, C_1, C_2 in Fig. 15.c) should be representative for the respective sites in a long chain molecule as far as

quantum chemistry is concerned. Endgroup effects (on the chemical shift of the central atoms) should be negligible. As the γ-*gauche* effect is known to be important, at least the γ-neighbors of the central atoms should be included in the model molecule. Moreover, these γ-neighbors should not be terminal atoms, as a terminal bonding situation is different from an atom inside the chain.

The model molecule in Fig. 15.c meets both requirements. It is comparatively small, and the γ-neighbors of the atoms A, B_1, B_2, C_1, C_2 for which chemical shifts will be calculated[1] are situated well inside the molecule.

The geometry of the model molecule is largely specified by the configuration of the central tetrad (x_1, x_2, x_3) and the conformation of the four dihedral angles $(\vartheta_1, \vartheta_2, \vartheta_3, \vartheta_4)$. If these parameters are set to well-defined values, the positions of the γ-neighbors of the central methylene unit A and the two neighboring methyl units C_1, C_2 are fixed; this region of well-defined geometry is depicted by bold lines in Fig. 15.c. Note that for the γ-neighbors of the tertiary carbons B_1 and B_2, precise values for ϑ_0 and ϑ_5 are required. To this end, a conformational statistics $p_6(\vartheta_0, \ldots, \vartheta_5 \mid x_1, x_2, x_3)$ is needed with a very large number of relevant conformations. As the methine carbons display quite narrow resonances in the experimental spectrum and are thus not of prime interest, ϑ_0 and ϑ_5 are left free and are set to arbitrary values in the course of the geometry optimization. For this reason, the simulation of the methine resonance will yield only tentative and approximate results.

The geometries themselves are generated by a force field minimization with the widespread Consistent Valence Force Field (CVFF) (Sect. 2.3) of Biosym Inc. [33]. The dihedral angles ϑ_i are forced to their selected values ϑ_i^j by an additional potential

$$V_{constrain} = \frac{1}{2} k_{constrain} (\vartheta_i - \vartheta_i^j)^2 . \tag{3.14}$$

The overall energy (force field energy V_{CVFF} plus forcing potential $V_{constrain}$) is minimized with respect to the nuclear coordinates of the molecule. The configuration is controlled during the minimization, and the other dihedral angles (including ϑ_0 and ϑ_5) are left free. As the central dihedral angles are essentially fixed to their RIS values, the energy dependence of the remaining degrees of freedom is well-behaved, and the geometries were optimized by consecutively employing the conjugated gradient and Newton-Raphson algorithm.

By this procedure, for each of the relevant conformations $k^j = (\vartheta_1^j, \vartheta_2^j, \vartheta_3^j, \vartheta_4^j; x_1^j, x_2^j, x_3^j)$ an optimized geometry $G^j = G(k^j)$ is obtained.

---

[1] In fact, the IGLO method provides the chemical shift tensors for all atoms in a single calculation, but only the values of the central atoms will be used for the simulation.

These optimized geometries are considered as representative of the geometries of the extended polymer chain. Afterwards, for each of the geometries G^j, an IGLO calculation is performed, which yields the ^{13}C chemical shifts for all nuclei K and all geometries G^j: $\sigma_K(G^j)$. The atomic basis DZ (see Sect. 2.4.2) is used throughout. The CPU time (quantum chemistry) for a single geometry is $\approx$ 12 h on a SGI IRIS INDIGO (CPU MIPS R4000, 64 MByte RAM), and 4 h–6 h on DEC ALPHA AXP 400/600/800 (175/200 MHZ, with at least 64 MByte RAM). The semidirect version of the IGLO program [41] was used, and routinely about 50 million integrals were stored on hard disk.

3.3.3
Simulated Spectrum as a Combination of Statistical Model and ab initio Quantum Chemistry

The statistics of the long polymer chain (that is inaccessible to explicit quantum chemistry calculations) is now transferred to the small model molecule (that can be conveniently treated on a workstation). Then the simulated line shape for the nucleus K is

$$g_K(\sigma) = \sum_j \underbrace{p_{config}(x_1^j, x_2^j, x_3^j)}_{\text{synthetic conditions}} \cdot \underbrace{p_{conform}^{RIS}(\vartheta_1, \vartheta_2, \vartheta_3, \vartheta_4 \mid x_1, x_2, x_3)}_{\text{statistical model}} \cdot \underbrace{\delta(\sigma - \sigma_K(G^j))}_{\text{IGLO}}$$

(3.15)

In this formula, p_{config} is the experimental configurational statistics of the sample to be simulated; it is fully determined by the synthetic conditions and is not subject to change. For ideally atactic poly(propylene), the configurational statistics is easily calculated, $p_{config}(x_1, x_2, x_3) = (\frac{1}{2})^3 = 0.125$, $x_i \in \{m, r\}$. $p_{conform}^{RIS}(\vartheta_1^j, \vartheta_2^j, \vartheta_3^j, \vartheta_4^j \mid x_1^j, x_2^j, x_3^j)$ is the conditional conformational statistics (Eq. 3.12) obtained from the RIS model.

The structure of Eq. (3.15) emphasizes the modular character of the present approach (cf. Fig. 6). The statistical model ([86] in this case), the way of generating the geometries G^j (here: CVFF empirical force field), and the quantum chemical method for the computation of the chemical shift (here: Hartree-Fock level IGLO scheme) are building blocks that can be exchanged, if necessary. More refined statistical models, better geometries e.g. by SCF-MP (Self Consistent Field with Møller-Plesset correlation corrections [45, 90]) and more refined ways to obtain the chemical shift can be used alike in Eq. (3.15) in order to improve the present methods. In contrast to this, the discrete nature of the sum in Eq. (3.15) with only comparatively few terms is essential in handling the problem with present day computing power. The incorporation of a quasi continuous probability distribution is virtually unattainable with the available means and methods.

To allow for deviations from the ideal RIS dihedral values, we replace the δ-function in Eq. (3.15) by a Gaussian function $s_b(x) = \frac{1}{\sqrt{2\pi}b} \exp(-x^2/2b^2)$.

$$g_K(\sigma) = \sum_j \underbrace{p_{config}(x_1^j, x_2^j, x_3^j)}_{\text{synthetic conditions}} \cdot \underbrace{p_{conform}^{RIS}(\vartheta_1^j, \ldots, \vartheta_4^j \mid x_1^j, x_2^j, x_3^j)}_{\text{statistical model}} \cdot \underbrace{s_b(\sigma - \sigma_K(G^j))}_{\text{IGLO}}$$

$$(3.16)$$

The width b of the Gaussian function was calibrated in separate calculations by deliberately changing the dihedral angles from their RIS values by $\pm 10°$ (a value suggested by other simulations [89, 91]) and monitoring the changes in the chemical shift. The deviations from the ideal dihedral value lead to variations of the chemical shift ranging from 0.2 ppm to 2.7 ppm for different geometries. To avoid the introduction of many numerical factors, an effective width of $b = 1.5$ ppm is assumed throughout this review for all groups and all polymers. This is certainly only a rough approximation of the "real" patterns but necessitated by today's computer resources. There are some indications (*vide infra* Sect. 5.1) that the chemical shift of a molecule in a *trans*-state is less sensitive to geometry variations than molecules in a *gauche* state, but these are second-order effects that are not accounted for in the idealized formula (3.16), but could readily be incorporated.

Note that only the nuclei $K = A, B_1, B_2, C_1, C_2$ in the central part of the model molecule can be considered representative of an extended polymer chain, as only here the geometry is well defined and the end-group effects can be neglected. As already mentioned, the positions of the γ-neighbors of the methine groups B_1, B_2 are not fully determined (lacking a definite value for ϑ_0 and ϑ_5), so the simulation for these tertiary carbons will be only indicative of a trend.

In the same spirit, the configurational splitting in solution can be calculated. The position of the configuration (x_1, x_2, x_3) is simply the conformationally averaged shift.

$$\sigma_K(x_1, x_2, x_3) = \sum_{j, \, x_1^j = x_1, x_2^j = x_2, x_3^j = x_3} \underbrace{p_{conform}^{RIS}(\vartheta_1^j, \vartheta_2^j, \vartheta_3^j, \vartheta_4^j \mid x_1, x_2, x_3)}_{\text{statistical model}} \cdot \underbrace{\sigma_K(G^j)}_{\text{IGLO}} .$$

$$(3.17)$$

In Eq. (3.17), the sum extends only over those geometries with the correct configuration (x_1, x_2, x_3). It should be noted, that by Eq. (3.17), the configurational splitting in solution may be computed without special assumptions on the geometry dependence of the chemical shift. Only if a simple functional dependence for $\sigma_K(G^j)$ is postulated, the phenomenological parametrisation of the γ-*gauche* effect is recovered (Sect. 2.5).

For the methylene carbon A, such a simplified dependence may be expressed as the sum of constant contribution σ_{abs} and geometry-dependent contribution γ_{CH_2} that is effective only if either ϑ_1 or ϑ_2 is in a *gauche* state

(The occurrence of several *gauche* states is neglected for the time being).

$$\sigma_A(G^j) = \sigma_{abs} + \gamma_{CH_2} \cdot \delta_{\vartheta_1,'gauche'} + \gamma_{CH_2} \cdot \delta_{\vartheta_4,'gauche'} \tag{3.18}$$

If Eq. (3.18) is inserted into Eq. (3.17), the phenomenological parametrization (see Sect. 2.5) emerges.

$$\sigma_A(x_1, x_2, x_3) = \sigma_{abs} + \gamma_{CH_2} \cdot P_{gauche,\vartheta_1}(x_1, x_2, x_3) + \gamma_{CH_2} \cdot P_{gauche,\vartheta_4}(x_1, x_2, x_3) \tag{3.19}$$

Here, $P_{gauche,\vartheta_i}(x_1, x_2, x_3)$ is the probability that ϑ_i is in a *gauche* state, and γ_{CH_2} is now an empirical constant. The probability P_{gauche,ϑ_i} depends on the configuration sequence (x_1, x_2, x_3) or even longer configurational sequences $(x_0, x_1, x_2, x_3, x_4)$, etc. The influence of long configurational sequences found in solution NMR is not at variance with the strictly local character of the chemical shift, as a remote asymmetric carbon atom acts by changing the probability rather than the chemical shift of a given conformation. By introducing more γ-*gauche* constants, the occurrence of multiple *trans* and *gauche* states in an RIS model may be taken into account.

It is now evident that the phenomenological formula is simply a parametrization relying on the stated geometry dependence Eq. (3.18). By quantum chemistry calculations, this assumption may be checked and compared to the predictions of Eq. (3.17).

3.3.4
Symmetry

To further reduce the necessary computing time, the intrinsic symmetry of the model molecule can be exploited. The two geometries $G(k)$ and $\overline{G} = G(\bar{k})$ determined by the sequence $k = (\vartheta_1, \vartheta_2, \vartheta_3, \vartheta_4; x_1, x_2, x_3)$ and its reversed counterpart $\bar{k} = (\vartheta_4, \vartheta_3, \vartheta_2, \vartheta_1; x_3, x_2, x_1)$ can be transformed one into another by rotation and reflection. The names of the atoms change accordingly. For example, the atom C_1 in geometry G is transformed into C_2 in $\overline{G}$. The chemical shift depends only on the relative positions of the atoms and is invariant under these symmetry operations. Consequently, the following relations should hold

$$\sigma_A(G(k)) = \sigma_A(G(\bar{k})) \tag{3.20}$$

$$\sigma_{B_1}(G(k)) = \sigma_{B_2}(G(\bar{k})) \tag{3.21}$$

$$\sigma_{C_1}(G(k)) = \sigma_{C_2}(G(\bar{k})) \tag{3.22}$$

In practice, in our calculation, small deviations (< 1 ppm for A, $C_{1,2}$) are observed as the symmetry is slightly violated by the non-defined exterior dihedral angles (e.g. ϑ_0, ϑ_5).

Using the symmetry relations Eqs. (3.20) -(3.22), and the corresponding expression for the RIS probabilities, Eq. (3.13), the IGLO calculation of the symmetry-related geometries $\overline{G}$ can be avoided, and the following formulae for the simulated solid state spectrum are obtained.

$$g_{CH_2}(\sigma) = \sum_j [p(k^j) \cdot s_b(\sigma - \sigma_A(G(k^j)))] \delta_{k^j,\bar{k}^j}$$

$$+ [2p(k^j) \cdot s_b(\sigma - \sigma_A(G(k^j)))](1 - \delta_{k^j,\bar{k}^j}) \tag{3.23}$$

$$g_{CH_3}(\sigma) = \sum_j [\frac{1}{2} p(k^j) \cdot (s_b(\sigma - \sigma_{C_1}(G(k^j))) + s_b(\sigma - \sigma_{C_2}(G(k^j))))] \delta_{k^j,\bar{k}^j}$$

$$+ [p(k^j) \cdot (s_b(\sigma - \sigma_{C_1}(G(k^j))) + s_b(\sigma - \sigma_{C_2}(G(k^j))))](1 - \delta_{k^j,\bar{k}^j}) \tag{3.24}$$

$$g_{CH}(\sigma) = \sum_j [\frac{1}{2} p(k^j) \cdot (s_b(\sigma - \sigma_{B_1}(G(k^j))) + s_b(\sigma - \sigma_{B_2}(G(k^j))))] \delta_{k^j,\bar{k}^j}$$

$$+ [p(k^j) \cdot (s_b(\sigma - \sigma_{B_1}(G(k^j))) + s_b(\sigma - \sigma_{B_2}(G(k^j))))](1 - \delta_{k^j,\bar{k}^j}) \tag{3.25}$$

In these formulae, $\delta_{k,k'}$ is a generalized Kronecker symbol,

$$\delta_{k,k'} = \delta_{\vartheta_1,\vartheta_1'} \cdot \delta_{\vartheta_2,\vartheta_2'} \cdot \delta_{\vartheta_3,\vartheta_3'} \cdot \delta_{\vartheta_4,\vartheta_4'} \cdot \delta_{x_1,x_1'} \cdot \delta_{x_2,x_2'} \cdot \delta_{x_3,x_3'} \tag{3.26}$$

and the sum extends only over conformations not related by the symmetry relation $k \leftrightarrow \bar{k}$. For the configurational splitting, Eq. (3.17), the symmetry can be used in a similar manner.

By applying these relations, the necessary computer time can be considerably reduced. Now only 90 geometries (instead of 140 without the use of symmetry) must be calculated explicitly. A halving cannot be achieved, as there are conformations that are invariant under reversal, $k = \bar{k}$.

3.3.5
Summary of the Approach

The ab initio simulation of solid state ^{13}C NMR spectra of synthetic polymers is performed in 4 steps (see Fig. 6).

1. Calculation of the conformational statistics $p_{conform}^{RIS}(\vartheta_1, \vartheta_2, \vartheta_3, \vartheta_4 \mid x_1, x_2, x_3)$ using an RIS model; identification of the important conformations with high probability.
2. Generation of the coordinates of all atoms in a small model molecule by force-field minimization with constrained dihedral angles.
3. Calculation of the chemical shift for the central atoms in each geometry by IGLO.

4. Combination of the results of the IGLO calculations with conformational probabilities, simulation of spectra by Eq. (3.16).

3.4
Simulated Solid State Spectrum – Comparison with Experiment

In Fig. 16, the simulated chemical shifts for the various groups are compared with experimental solid state spectra. As already mentioned, the experimental spectrum was recorded well below T_g, so motional effects affecting the line shape can be excluded. The bold lines in the simulated spectra display the overall simulation (according to Eq. (3.16)) whereas the fine lines represent the contributions of the various conformations (Eq. (3.15)) which have been slightly broadened for typographical reasons. The width and the shape of the methylene resonance are well reproduced; the two main contributions and the small upfield shoulder show up both in the experimental and in the theoretical spectrum. The fine lines underneath the overall simulation reveal that a multitude of different geometries with different chemical shifts do contribute and are relevant for the observed line shape. From this finding it can already be concluded that the naive *γ-gauche* effect must be modified to explain this pattern. If the chemical shift depended only on the traditional "*γ-gauche* dihedral angles", only a few sharp contributions should be discernible. We shall investigate this point in some more detail (*vide infra*, Sect. 3.5). This will also allow the identification and assignment of the various subresonances.

The methyl unit is comparably well met; the experimental features – main contribution around ≈ 19 ppm and large upfield shoulder – are clearly visible in the simulated spectrum, too. The methine resonance is quite narrow both in the experiment and the simulation; moreover, a small but extended downfield shoulder is reproduced. It should be recalled, however, that the position of one *γ*-neighbor of the methine carbons investigated (B_1, B_2 in Fig. 15.c) was not specified in the course of the simulation. Thus, the theoretical results are only indicative of a trend.

The good overall agreement of experiment and theory refers to *shape* and *width* and not to absolute chemical shift values. In fact, the absolute values of the chemical shift are not properly reproduced due to the small atomic basis employed (*vide infra*, Sect. 4.3). Thus, the position of the simulated spectrum was adjusted for each resonance separately.

The *shape* of the bands in the spectrum is given by differences of the shift of the same nucleus within the same molecule, but different geometries, viz.

$$\Delta\delta_X = \delta_X(G^j) - \delta_X(G^{j'}) . \tag{3.27}$$

In this expression, errors in the absolute value (which are due to the small atomic basis) cancel almost completely. Consequently, for our purposes, the

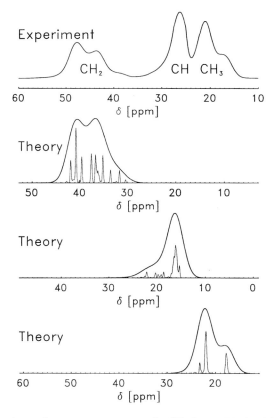

Fig. 16. The experimental CP/MAS spectrum of solid glassy aPP (*top*) is compared with simulations for the various groups (*rows 2-4*). For all groups the shape is satisfactorily matched. Note that for all spectra – experiment and theory – the width of the spectrum is 50 ppm. The absolute position is not correctly reproduced by the quantum chemical simulations due to the small atomic basis employed and has been adjusted for better comparison. The absolute height of the peaks does not convey any information since, experimentally, different cross polarization efficiencies tend to distort the stoicheometric relations. Thus, the simulated spectra were all scaled to the same peak height. The *fine line* shows the contributions of the various conformations, Eq. (3.15) (slightly broadened for typographical reasons), the *bold line* is the simulation with the δ-function being replaced by a Gaussian function s_b, Eq. (3.16)

small atomic basis is sufficient. On the other hand, for the prediction of the solution spectrum of a newly synthesized compound, the absolute value of the chemical shift matters, and larger atomic basis sets must be adopted. A more detailed account of this topic will be given in Sect. 4.3.

For a similar reason, the heights of the various subresonances for a given group are well reproduced by the simulation method – the relative heights reflect the statistical probability for the occurrence of the underlying geometries. But the intensity of a resonance in an experimental solid state NMR spectrum depends also on the cross polarization efficiencies and other experimental uncertainties. Naturally, these experimental and technical effects cannot be accounted for by a simulation which is based exclusively on statistical mechanics and quantum chemistry.

3.5
Correlation of Chemical Shift and Geometry – the γ-*gauche* Effect Revisited

The theoretical investigation of inhomogeneously broadened ^{13}C -NMR spectra provides the possibility to correlate the chemical shift with geometrical parameters. As a first application, the influence of the dihedral angles ϑ_1 and ϑ_4 on the methylene chemical shift is depicted in Fig. 17. The overall decrease of δ with increasing $|\vartheta_1| + |\vartheta_4|$ reflects the phenomenological γ-*gauche* effect. Thus, the basic empirical finding is reproduced by theoretical investigations. On the other hand, for a fixed value of $|\vartheta_1| + |\vartheta_4|$ a considerable spread in δ is observed (horizontally aligned clusters). This spread is due to non-γ-*gauche* contributions, and the figure shows that the phenomenological understanding of a γ-*gauche* effect depending on ϑ_1 and ϑ_4 alone, is largely oversimplified. Neither the occurrence of various clus-

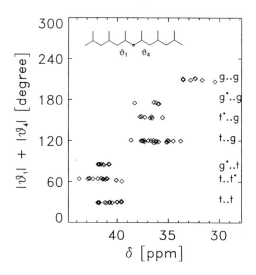

Fig. 17. The methylene group in aPP displays a γ-*gauche* effect as the chemical shift δ is correlated with $|\vartheta_1| + |\vartheta_4|$. The seven horizontally aligned clusters stem from different combinations of the rotational isomeric states. The considerable spread within the clusters (with essentially the same value of $|\vartheta_1| + |\vartheta_4|$) betrays the influence of other geometric parameters which are not accounted for in the simple phenomenological understanding of the γ-*gauche* effect

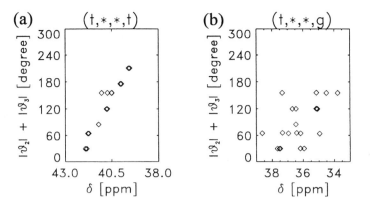

Fig. 18. For the cluster belonging to ($\vartheta_1 = t, \vartheta_2 = *, \vartheta_3 = *, \vartheta_4 = t$), the chemical shift of the methylene residue is correlated with $| \vartheta_2 | + | \vartheta_3 |$ (a). This shows that next to the exterior dihedral angles ϑ_1 and ϑ_4, the interior dihedral angles ϑ_2, ϑ_3 are relevant for the chemical shift. For other clusters like ($\vartheta_1 = t, \vartheta_2 = *, \vartheta_3 = *, \vartheta_4 = g$), there is no apparent pattern for the observed spread of the chemical shifts within a cluster (b)

ters due to multiple *trans-* and *gauche*-states nor the spread within the clusters are taken into account in the simple parametrization, Eq. (2.47), (2.48).

Apart from ϑ_1, ϑ_4, other variables have an impact on the chemical shift. In fortunate cases, these additional variables may be identified. In Fig. 18, some clusters with fixed $| \vartheta_1 | + | \vartheta_4 |$ are further analyzed and the chemical shift δ is plotted vs. $| \vartheta_2 | + | \vartheta_3 |$ for each cluster separately. For the cluster $| \vartheta_1 | + | \vartheta_4 | \approx 30°$ (i.e. $\vartheta_1 = \vartheta_4 = t$, ϑ_2, ϑ_3 not specified), the remaining spread in δ is strongly correlated with $| \vartheta_2 | + | \vartheta_3 |$. Thus, next to the "classical" γ-*gauche* dihedral angles ϑ_1, ϑ_4, the interior angles ϑ_2, ϑ_3 influence the chemical shift. The pattern, however, is mostly more complicated than just a linear correlation with $| \vartheta_2 | + | \vartheta_3 |$, as can be seen in Fig. 18.b, where the cluster ($\vartheta_1 = t, *, *, \vartheta_4 = g$) is investigated. The asterisk denotes an unspecified dihedral angle. For the very simple system poly(ethylene), the influence of the various geometric parameters can be completely disentangled. Again, the interior dihedral angles are almost as important as the exterior "γ-*gauche* angles" (*vide infra*, Sect. 5.1).

For the methyl carbon C_1, the γ-*gauche* sensitive dihedral angles are ϑ_1 and ϑ_2 (Fig. 19.a). Note that a *trans* value of the main chain dihedral angle ϑ_2 implies that the methine unit B_2 is *gauche* to the methyl group C_1 in question (Fig. 19.b). Consequently, the influence of $| \vartheta_1 | + | \vartheta_2 |$ on δ should be reversed compared to Fig. 17. This is exactly what is displayed in Fig. 19.c, where a very neat correlation of δ_{CH_3} with $| \vartheta_1 | + | \vartheta_2 |$ is observed. Again, the γ-*gauche* effect is essentially reproduced by the quantum chemical

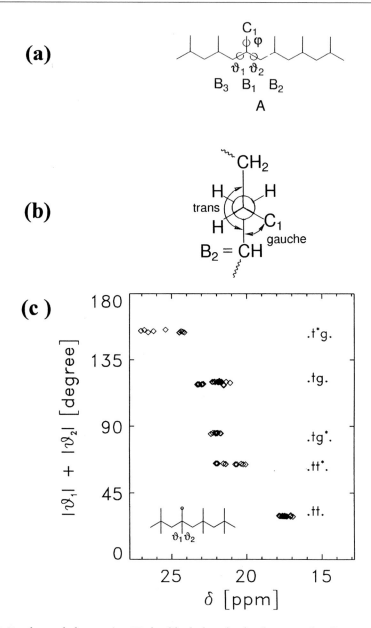

Fig. 19. For the methyl group in aPP, the dihedral angles for the γ-*gauche* effect are ϑ_1 and ϑ_2; the "interior" dihedral angle φ simply parametrizes the rotation of the methyl group C_1 (**a**). A *trans* value of the main chain dihedral angle ϑ_2 implies that the γ-neighbor B_2 is in a *gauche* position with respect to the methyl group C_1 (**b**). For this reason, the chemical shift δ_{CH_3} decreases with a decreasing value of $|\vartheta_1| + |\vartheta_2|$ (**c**). The correlation is much better than for the methylene group (Fig. 17), and the spread within the horizontal clusters is much smaller

calculations. Compared with Fig. 17, the observed spread within the clusters (of fixed $| \vartheta_1 | + | \vartheta_2 |$) is significantly reduced. This finding further points to the importance of the internal dihedral angles. For the methyl group, the interior dihedral angle φ (Fig. 19.a) simply parametrizes the rotation of the methyl group hydrogens and has thus very little importance for overall geometry. This is why the spread within the clusters of the methyl chemical shift is reduced. This assertion does not hold for the methylene unit, where the overall shape of the molecular fragment depends on both the exterior angles ϑ_1, ϑ_4 and the interior dihedral angles ϑ_2, ϑ_3.

The γ-*gauche* effect is only partly steric in origin (Fig. 20) since the combined distances of the γ-neighbors do not fully explain the calculated chemical shift; the correlation of δ with the γ-neighbor distances is far from being perfect.

One of the tacit assumptions of the phenomenological parametrization of the γ-*gauche* effect is the strictly tetrahedral coordination of the carbon atoms involved. In reality, the bond angle ψ depends on the two neighboring dihedral angles and widens if these change from a *trans* to a *gauche* position (Fig. 21.a). As a consequence, there is even a correlation of the chemical shift with this bond angle (Fig. 21.b). Such correlations should not be overemphasized. They occur because the various geometric parameters are mutually interrelated. Thus a correlation of the first geometric parameter with δ implies a relation of the chemical shift and the second. In some cases, the apparent dependence of the chemical shift upon such a geometrical parameter may be astonishingly good (see the case of poly(isobutylene), Sect. 5.2), but this *correlation* must not be confused with an *explanation*, a causal relation.

The Figs. 17 and 19 allow the assignment of geometries to the various subpeaks in the experimental spectrum. This assignment which is based exclusively on theoretical arguments is depicted in Fig. 22. In the fortunate case of poly(propylene), it may be verified by comparison with well-defined crystalline modifications of isotactic (*tg*-helix) and syndiotactic (*ttgg*-helix) poly(propylene) (see e.g. [8]). In crystalline iPP, there is only one type of methylene unit which has a *trans* and a *gauche* γ-partner, whereas in sPP there are two types with two *trans* and two *gauche* γ-neigbors, respectively. The experimental peak positions of the crystalline sample exactly coincide with the theoretical assigments (Fig. 23). The same holds true for the other resonances. The theoretical assignment is thus verified by experiment. Note that the upfield shoulder of the methyl unit cannot be assigned by comparison with crystalline samples but only by calculational methods. Seen from another viewpoint, the common assignment of subresonances in amorphous systems by comparison with crystalline samples is now substantiated by ab initio calculations.

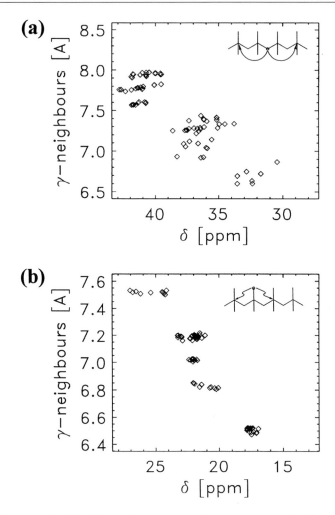

Fig. 20. The *γ-gauche* effect has an important steric component as the chemical shift is correlated with the distance of the carbon γ-neighbors for both the methylene (**a**) and the methyl (**b**) group. The considerable scatter in both plots shows that the geometry dependence of the chemical shift cannot be understood by steric reasons alone

To conclude, the ^{13}C chemical shift in NMR spectra of solid polymers evidently reflects *conformational* rather than *configurational* variety, as is sometimes stated in the literature. [92, p. 180] [8, p. 215] [93]

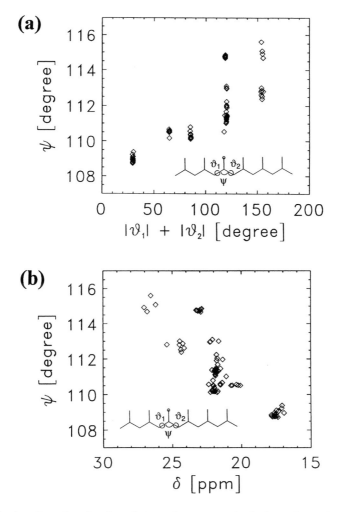

Fig. 21. The bonding situation in poly(propylene) strongly deviates from ideal tetrahedral coordination, as the bond angle ψ deviates from the textbook value $\psi = 109°$ and moreover depends on the dihedral angles ϑ_1 and ϑ_2 in a systematic fashion (**a**). As a consequence, the chemical shift is correlated with ψ (**b**)

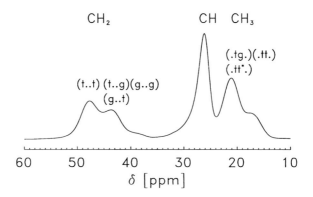

Fig. 22. By correlating the theoretical chemical shifts with the underlying geometries, the subpeaks of the experimental spectrum can be assigned. The *dots* symbolize unspecified dihedral angles

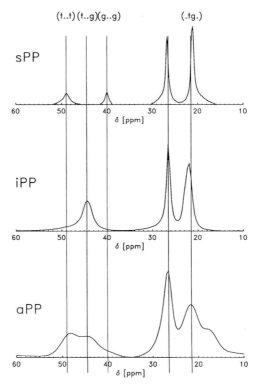

Fig. 23. In poly(propylene), the theoretical subpeak assignment can be checked by comparison with the NMR spectra of crystalline samples of syndiotactic (*top*) and isotactic (*middle*) poly(propylene). The methylene region of syndiotactic poly(propylene) consists of two peaks belonging to a (*t..t*) and a (*g..g*) conformational environment, respectively, whereas the single peak of iPP belongs to the (*t..g*) conformation. They perfectly match the subpeaks in aPP (*bottom*) and do confirm the theoretical assignment. The same holds true for the other resonances

3.6
Molecular Orbital Contributions to the Chemical Shift

The IGLO method permits the identification of localized molecular orbital contributions to the chemical shift that can be interpreted on a physico-chemical basis. Contributions, e.g. for carbon $1s$ core electrons, for C-H and C-C bonds, etc. can be identified. Even double and triple bonds can be described. In this section, the MO contributions to the methylene carbon chemical shift will be investigated (Fig. 24). For simplicity, in the following, we will refer to the theoretical scale σ rather than to the experimental scale δ which is measured with respect to a reference.

The following essential features are found with little variation for all the systems presented in this study.

- The contribution of the carbon $1s$ shell is dominant and does not vary with geometry ($\sigma_{MO_0} \approx 201$ ppm).
- The molecular orbitals of the adjacent bonds $MO_1, \ldots, MO_4$ contribute up to -10 ppm each and depend on the conformation. The γ-gauche effect and the other geometry effects show up in these molecular orbitals.
- The other localized molecular orbitals are fairly unimportant (< 1 ppm each). Among these small contributions, the γ-bonding orbitals (dashed lines in Fig. 24) are the most important. Their contribution does not exceed 0.9 ppm and does depend on geometry. This is another consequence of the interaction of an atom with its γ-neighbors.

These findings show that NMR is indeed a strictly local method; the chemical shift of a nucleus is determined by the electrons in its immediate vicinity. The influence of remote residues is fairly small and acts mainly by changing the neighboring MOs. The direct contribution of distant molecular orbitals is negligible.

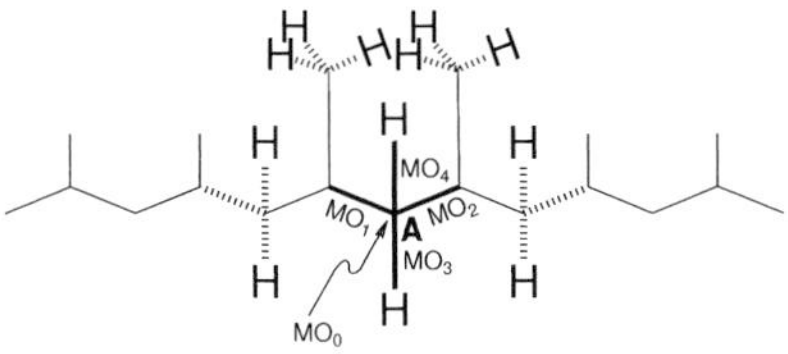

Fig. 24. The chemical shift is a strictly local phenomenon. The main MO contributions of the chemical shift of the central methylene carbon stem from the carbon $1s$ core electron MO_0 and the MOs of the four adjacent bonds $MO_1, \ldots, MO_4$. The latter depend on the conformation. The influence of remote MOs is small. Among these, the γ-MOs (denoted by a *dashed line*) are the most important, reflecting the interaction of the carbon atom with its γ-neighbors

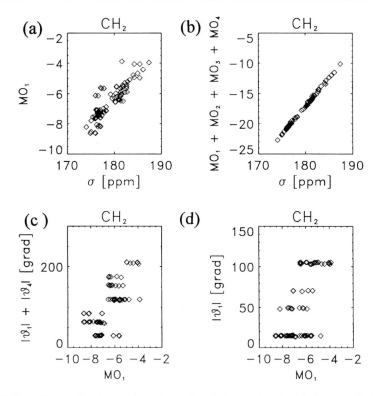

Fig. 25. The MO contribution of the adjacent bonds is correlated with the overall chemical shift (**a**). The four adjacent-bond MOs are responsible for almost all of the geometry-induced variation of the chemical shift (**b**). The MOs display a γ-*gauche* effect separately (**c**), although there is a slightly better correlation of MO_1 with the corresponding dihedral angle ϑ_1 (**d**)

The geometry sensitive MOs MO_1, MO_2, MO_3, and MO_4 are strongly correlated with the overall chemical shift (Fig. 25.a, MO_1 as an example) and consequently with one another. The sum $\sigma_{MO_1} + \sigma_{MO_2} + \sigma_{MO_3} + \sigma_{MO_4}$ is sufficient to explain almost all of the observed spread in the chemical shift (Fig. 25.b), i.e. the geometry dependence is essentially confined to these four molecular orbitals. Because of the correlation of the separate MO contributions with the total chemical shift, all of them display to a certain degree the same geometry dependence (Fig. 25.c, for the γ-*gauche* effect). It is impossible to map a geometry parameter on "its" MO. A variation of a parameter always affects all orbitals, though the influence may be more or less pronounced. In this sense, the correlation of σ_{MO_1} with ϑ_1 (Fig. 25.d) is slightly better than the correlation with $| \vartheta_1 | + | \vartheta_4 |$. The molecular orbitals thus have a certain tendency to reflect to geometric situation of "their" side of the molecule. But apart from this rather small effect, there does not seem to be a convincing relation of molecular orbital and geometric parameter. Changing the geometry affects the molecule as a whole and all its major molecular orbitals.

For the methyl and methine carbons, the situation is essentially the same, and no new aspects occur.

3.7
Configurational Splitting in Solution

The ab initio IGLO calculations can be used to calculate the configurational splitting in solution by simply calculating the conformational average for a given configuration (x_1, x_2, x_3). For the reader's convenience, the formula Eq. (3.17) derived in Sect. 3.3.3 is adopted to the special case of the methylene unit A of the model molecule

$$\sigma_A(x_1, x_2, x_3) = \sum_{j,\, x_1^j = x_1, x_2^j = x_2, x_3^j = x_3} p_{conform}^{RIS}(\vartheta_1^j, \vartheta_2^j, \vartheta_3^j, \vartheta_4^j \mid x_1, x_2, x_3) \cdot \sigma_A(G^j) .$$

(3.28)

The sum is over the geometries with given configuration (x_1, x_2, x_3). In practice, only the conformations with significant probability are taken into account. As for the solid state NMR spectrum, symmetry is used to reduce the number of geometries to be calculated. The conformational statistics $p_{conform}^{RIS}(\vartheta_1^j, \vartheta_2^j, \vartheta_3^j, \vartheta_4^j \mid x_1, x_2, x_3)$ is recomputed for the necessary temperatures. Due to the small size of the molecule, only tetrad configurational splittings (x_1, x_2, x_3) of the methylene unit can be investigated. Experimentally, hexad splittings $(x_0, x_1, x_2, x_3, x_4)$ can be resolved. To allow a comparison with theory, the experimental values were averaged over the leftmost and rightmost configurational degree of freedom (x_0, x_4). In Fig. 26, the simulated configurational splittings (■) of the methylene group are compared with experimental data [94, 85] (averaged over x_0, x_4) for two temperatures. The predictions of the phenomenological method [94] are also shown (□). Only the configurational splitting $\Delta\delta$ with respect to some arbitrary reference is given. At $T = 315$ K, the experimental findings are well reproduced by both the ab initio calculations and the empirical method; at elevated temperatures, however, the phenomenological approach seems to yield more accurate results. The higher numerical accuracy of the empirical method has two causes.

1. At higher temperatures, the number of relevant conformations increases sharply because of the underlying Boltzmann statistics. The overall probability of the ≈ 90 geometries used in the quantum chemistry calculations is $p \approx 0.85$ at $T = 315$ K compared with only $p \approx 0.75$ at $T = 391$ K. Computer time considerations impede the ab initio treatment of many

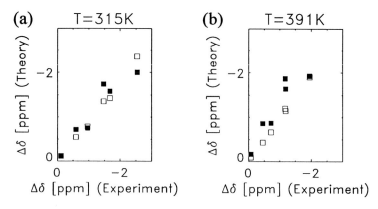

Fig. 26. Configurational splitting for the methylene position in poly(propylene) at $T = 315$ K (**a**) and $T = 391$ K (**b**). The predictions of the IGLO calculations (■) and the empirical γ-*gauche* method (□) are plotted vs. experimental results. The experimental results and the empirical predictions were taken from [94], where hexad assignments $(x_0, x_1, x_2, x_3, x_4)$ are given. For a better comparison with the IGLO results (tetrad resolution) an average over the exterior diads x_0 and x_4 has been performed. The theoretical predictions of the IGLO method are better at the lower temperature

hundred geometries, whereas the phenomenological approach takes into account all conformations.

2. Within certain limits, the γ-*gauche* constants γ_i can be adjusted to the given situation, and in [94] the constant describing the impact of CH on CH_2 was adapted to a special value. Such a fitting procedure naturally improves the overall agreement of experiment and calculation.

Considering present day computing power, the ab initio method should not be considered as a convenient and fast standard method to obtain configurational splittings, but rather as a theoretical check of the empirical scheme. Figure 26 clearly shows, that for poly(propylene) at least, the *ad hoc* assumption Eq. (3.18) for the geometry dependence of the chemical shift is sufficient for the prediction of the configurational splitting. The phenomenological method is more accurate at higher temperatures, provides a better resolution (hexads), and needs very little computer time. It is thus the first candidate for a *rapid* assignment of the configurational splitting. These obvious advantages are based, however, on the occurrence of a number of poorly known parameters (the γ-*gauche* constants) and a special functional dependence of the chemical shift (Eq. 3.18). The ab initio method, on the other hand, does not rely on free parameters and special assumptions. For systems, where the geometry dependence of the chemical shift is more complicated, the phenomenological approach is bound to fail whereas the quantum chem-

Table 5. Configurational splitting of the methyl carbons in atactic poly(propylene). Because of the small size of the model molecule, only triads can be resolved. The reference for the splitting is the *mm* configuration

Triads	[94] $T = 315$ K	IGLO $T = 315$ K	[94] $T = 391$ K	IGLO $T = 391$ K	[85] $T = 373$ K
mm	0.0	0.0	0.0	0.0	0.0
mr/rm	−0.646	−0.53	−0.68	−0.59	−0.73 / −0.77
rr	−1.506	−1.11	−1.45	−1.12	−1.40 / −1.53

ical scheme should work irrespective of the nature of the geometry effects. An example will be presented in Sect. 5.5.

For the methyl unit, only triad splittings have been calculated. The results are shown in Table 5; the findings do not differ much from those already presented. The splitting of the methine carbon is minute and will not be discussed.

It should be kept in mind, that the calculation of the configurational splitting either by the empirical method or the ab initio technique relies on the absence of solvent effects. For poly(propylene), this tacit assumption is fulfilled [94].

3.8
The Anisotropy of the Chemical Shift as a Source of Structural Information

The chemical shift is a tensorial quantity, i.e. it is anisotropic and depends on the orientation of the molecule with respect to the external magnetic field.

$$\sigma = \sigma_{iso} + \frac{1}{2}\delta_{aniso}[(3\cos^2\vartheta - 1) - \eta\sin^2\vartheta\cos(2\phi)] \qquad (3.29)$$

In this formula, (ϑ, ϕ) are the spherical angles of the magnetic field in the principal axes system of the symmetrized chemical shift tensor (Sect. 2.4.3) $\sigma_{\alpha\beta} = \sigma_{\beta\alpha}$; σ_{iso} denotes the isotropic shift, δ_{aniso} is the anisotropy parameter, and η the asymmetry parameter. For a detailed discussion of the definitions and conventions we refer to the literature, e.g. [3]. In a static 1D NMR experiment of a simple, well-defined powder sample, typical powder patterns are recorded (Fig. 27). The singularities are easily interpreted as the principal components $\sigma_{xx}, \sigma_{yy}, \sigma_{zz}$ of the diagonalized shift tensor. In systems with multiple resonances or inhomogeneous broadening, the different powder patterns normally overlap and yield a complicated spectral distribution which conveys little information. Then, it is necessary to separate the distribution in the isotropic shift from the anisotropic powder pattern by special experimental techniques. A conceptually simple, but experimentally quite demanding approach is a **Dynamic Angle Spinning (DAS)** experiment [95, 96, 97] [3,

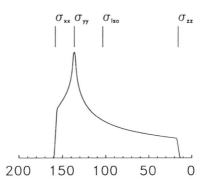

Fig. 27. Typical powder pattern reflecting the anisotropy of the chemical shift. The edges of the pattern and the maximum give the principal tensor components σ_{xx}, σ_{yy}, and σ_{zz} directly. In complicated disordered systems, different powder patterns overlap and the resulting spectrum is usually too complicated for an interpretation

Chapt. 6.4/6.5] which requires special equipment. In this two-dimensional sample spinning method, the angle of rotation with respect to the external magnetic field is different in the evolution and detection period, respectively (Fig. 28.a). The method relies on the fact, that a fast rotation around an axis with an angle Θ_R with respect to the external magnetic field results in the scaling of the powder pattern with a scaling factor

$$s(\Theta_R) = \frac{1}{2}(3\cos^2\Theta_R - 1) \tag{3.30}$$

In the evolution period, the angle of rotation matches the magic angle ($\vartheta_2 = \vartheta_m = 54.7°$), the scaling factor $s(\vartheta_m)$ vanishes, and the spin system develops isotropically whereas in the detection time, the anisotropy of the chemical shift is encoded ($\vartheta_1 \neq \vartheta_m \Rightarrow s(\vartheta_1) \neq 0$). For a more detailed description of the experiment we refer to the literature cited above.

Fortunately, the IGLO method provides the full chemical shift tensor, and not just the isotropic value. The contour plot in Fig. 28.b shows an isotropic/anisotropic separation spectrum of atactic polypropylene, which has been recorded at $T \approx 250$ K with a home-built DAS probehead [98] on a BRUKER MSL-300 spectrometer. For every isotropic shift on the x-axis, the intensity distribution along the y-axis displays the corresponding powder pattern; the edges correspond to σ_{xx} and σ_{zz} as indicated in the figure. The theoretical values of the xx- and the zz-component are superimposed on the experimental contour plot in Fig. 6 and symbolized by the small rhombi ($\diamond$). The absolute values were adjusted in the same way as for the isotropic chemical shift. The spectral position of σ_{yy} is not easily seen in the experimental spectrum; to keep the figure simple, the theoretical values for σ_{yy} have been

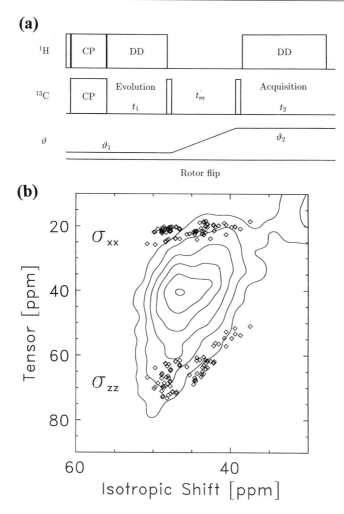

Fig. 28. Scheme for the 2D-DAS isotropic/anisotropic separation experiment (**a**). In the evolution period, the rotor spins around an axis with an angle ϑ_1 with respect to the external magnetic field. The magnetization is stored along the z-axis while the angle of rotation changes to ϑ_2, where the dectection takes place. The evolution and detection period are proton decoupled, transversal magnetization is created by the cross polarization technique.

Experimental spectrum of the methylene region of glassy atactic poly(propylene) (**b**) recorded at $T \approx 240$ K on a BRUKER MSL-300 with a DAS probehead. The two angles for evolution and detection, respectively, were $\vartheta_1 = 81.7°$ and $\vartheta_2 = \vartheta_m = 54.7°$. The rotation frequency was 3000 Hz, and 128 t_1-increments were recorded. The theoretical values ($\diamond$) are superimposed on the experimental contour plot. The absolute positions were adjusted like in the 1D spectrum. A qualitative agreement of theory and experiment is seen

Table 6. Tensor values for the two methylene positions in syndiotactic poly(propylene) calculated by IGLO. In this case, the X-ray geometry was used as basis for the IGLO calculations

Resonance	Conformation	δ_{xx}	δ_{yy}	δ_{zz}	δ_{iso}
exterior	(t, g, g, t)	11.49	46.33	58.94	38.92
interior	(g, t, t, g)	14.45	33.11	37.77	28.44

omitted. Due to experimental imperfections (mainly a poor S/N ratio and a small deviation of ϑ_2 from ϑ_m) only some basic features may be compared.

We find theoretically, that σ_{xx} is fairly constant whereas the variation of σ_{zz} is considerably larger ($\approx$ 20 ppm) and dominates the variations of the isotropic chemical shift. The same trend is observed in the experimental spectrum. As a consequence, the size of the anisotropy $| \sigma_{xx} - \sigma_{zz} |$ decreases with decreasing isotropic shift both in the simulation and the experiment.

The latter point is substantiated by a simulation of crystalline syndiotactic poly(propylene) [11] which was based on the experimental X-ray geometry. The overall anisotropy $| \sigma_{zz} - \sigma_{xx} |$ of the interior and exterior CH_2 was found to differ by about 15 ppm (Table 6) in good agreement with the simulations for the amorphous sample.

Thus, in addition to the isotropic shift, the anisotropic part of the shift tensor is susceptible to geometry variations and might be used as a source of structural information. For more precise and especially for quantitative statements, better experimental data are required.

4
Some Remarks on the Method

In this chapter additional methodical aspects of the ab initio simulations will be discussed, including requirements, accuracy, and limitations.

4.1
Intra- and Intermolecular Contributions to the Chemical Shift

The formula (Eq. (3.16)), for the computation of a solid state NMR spectrum silently assumes that the chemical shift is mainly an *intra*molecular phenomenon. The chemical shift is supposed to depend on the geometry of the molecule rather than on the environment and the interaction with other molecules (*inter*molecular packing effects). For poly(propylene), the intermolecular effects may be estimated. A first clue is given in Fig. 23, where the ^{13}C -NMR spectra for crystalline syndiotactic, crystalline isotactic, and glassy atactic poly(propylene) are shown. In all of these samples, there is a $(\vartheta_1, \vartheta_2) = (t, g)$ contribution to the methyl signal. Although the molecular environments differ strongly (two crystalline and one amorphous packing), the chemical shift for this signal is about the same with a deviation of less than 1 ppm. Similar findings hold for the methylene and methine region. Thus, variations of the intermolecular contribution do not seem to exceed $\approx$ 1 ppm.

There is even a more direct evidence for the negligibility of direct packing effects in the case of poly(propylene). For isotactic poly(propylene), there are various crystalline packings; and in α-iPP and β-iPP, the same 3_1 helix with a *tgtg*- sequence of conformations is packed in different ways within the crystal (Fig. 29, [8]). The corresponding CP/MAS spectra differ very little. In the α-form, a splitting of the CH_3 and CH_2 resonances is observed which is due to inequivalent atomic positions within the crystal. The splitting as well as the deviation from the spectral position in the β-form is less than $\approx$ 1 ppm. As the geometry of the chain is identical in the two cases (at least within the experimental resolution of an X-ray analysis), these small deviations are intermolecular in origin. The spread of $\approx$ 1 ppm is less than the width of the Gaussian broadening function in Eq. (3.16). Thus, packing effects of this size

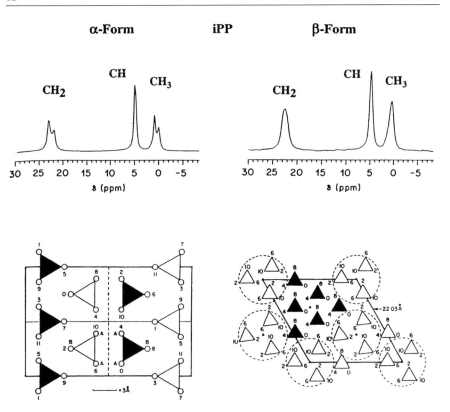

Fig. 29. The NMR spectra of two crystalline modifications of isotactic poly(propylene) are compared (*top row*). In the two modifications, the same *tg*-helix is packed in different ways (*bottom row*). As the position of the resonances differ only very slightly, the intermolecular (electronic) packing effects can be estimated to be small (< 1 ppm). Reprinted with minor modifications from [8]

are included in the previously stated uncertainties of the spectral simulation presented in this study.

It should be noted that the packing can influence the chemical shift by two different mechanisms. In the first, *direct* route, the different electronic environment causes a change in the chemical shift by modifying the wave function without affecting the geometry. On the other hand, the geometry of the molecule itself may be slightly distorted by the presence of its neighbors, and it is the modified geometry that causes the change in the chemical shift. This latter mechanism is an *indirect* packing effect. At present, the relative sizes of the two contributions are not clear.

The above discussion of poly(propylene) has targeted the direct electronic packing effects. The smallness and negligibility of these effects are a *conditio sine qua non* for the ab initio simulation of the NMR spectra of (amorphous) polymers. The explicit calculation of these electronic effects is impossible in practice, as a whole array of molecules in a multitude of different packings would have to be taken into account. Facing present day computing resources, the expenditure for such a procedure is prohibitive. In the following, the direct packing effects are assumed to be small for all of the fairly unpolar systems treated. Electronic packing effects might be important in polar samples or samples with hydrogen bonding [92, p. 180], but very recent investigations indicate that even for poly(vinyl alcohol) the intramolecular contributions to the chemical shift dominate [75].

The situation is different for the indirect packing effects. They can be calculated by the ab initio method if a structural model predicts the packing distortions with sufficient precision. Naturally, it will be hard to develop such a geometric-statistical model, but once the geometries are known they may be used as an input for the IGLO calculations and thus for the simulation of the ^{13}C -NMR spectrum.

For the simulation of the configurational splitting in solution, solvent effects must be virtually absent. This can easily be checked by recording the spectra with different solvents.

4.2
Influence of the Endgroups

To meet the available computer resources, the extended chain molecule has to be represented by a fairly small model molecule for the quantum chemistry calculations. It is to be demonstrated that no excessive errors are introduced by truncating the long polymer chain. In other words, the endgroup effects must be shown to be small. Again, there are two ways, remote parts of the chain molecule might influence the chemical shift in the central part.

1. The electronic wave function Ψ is modified by the endgroups although the geometry in the central part is unimpaired (direct effect).
2. The geometry of the central part of the molecule is slightly changed by the endgroups, and it is this perturbed geometry that affects the chemical shift (indirect effect).

In Fig. 30, three different geometries of the model molecule are shown. They have been minimized by fixing the four central dihedral angles, but the non-controlled dihedral angles (like ϑ_0 and ϑ_5) vary. The maximum deviation of the chemical shift (largest value - smallest value) is 1.86 ppm for the methyl group (C_1, C_2) and 2.00 ppm for the methylene group (A). The deviation of almost 2 ppm is quite large but still tolerable. This finding demonstrates that a molecule is sufficiently specified by the four central dihedral angles and that the uncontrolled relaxation of the additional degrees of freedom

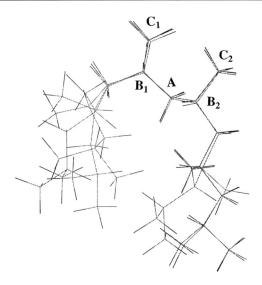

Fig. 30. Three geometries of the model molecule with identical central dihedral angles ϑ_1, ..., ϑ_4, but different values for the unspecified dihedral angles. The chemical shifts of the methyl and methylene resonances in the central part do not differ much. The smallness of endgroup effects is thus established

does not introduce large errors. A close inspection of the geometries reveals, that the geometry in the central part is slightly different in spite of the fixed dihedral angles ϑ_1, ..., ϑ_4. This is due to the geometrical influence of the endgroups. The error of ≈ 2 ppm is thus mainly due to the indirect route.

To assess the direct electronic influence, four already minimized geometries were selected, and their terminal isobutyl groups were replaced with hydrogens. Apart from this, no change in geometry was performed, and a renewed geometry optimization was avoided. The chemical shift of the truncated molecules differs from the shift of the original geometries by an average 0.31 ppm for CH_3 and 0.40 ppm for CH_2. Even the deviation for CH (which does not have a proper γ-partner in the truncated molecule) is very small (0.48 ppm). The direct electronic effect is thus very small – another argument for the strictly local nature of the chemical shift.

To conclude, the endgroup effects do not exceed 2 ppm (maximum - minimum value) and are still tolerable. The dominating contribution is the indirect geometry effect whereas the direct electronic effect is negligible. The main source of error is thus in the way of generating the geometries and not in the quantum chemistry calculations. This finding has two implications. First, special care is required to include all geometry effects and to use realistic geometries. Secondly, once the geometry is established, the quantum chemistry calculation may be performed on rather small sections of the chain molecules. This substantially reduces the required computation time.

Some more results on endgroup effects will be presented in Sect. 5.2.

4.3
Influence of the Atomic Basis and the Empirical Force Field

As already mentioned in Sect. 3.4 (see Fig. 16), the shape and the width of the experimental CP/MAS resonances are well reproduced by the simulation technique, whereas the absolute positions are less well met. The shape of a resonance is given by differences like

$$\Delta\delta_X = \delta_X(G^j) - \delta_X(G^{j'}), \qquad X = CH_3, CH_2, CH \qquad (4.1)$$

of the chemical shifts of the same site within the same molecule but in different conformations. Errors of the absolute shift cancel in Eq. (4.1). It should be noted, however, that the theoretical scale in Fig. 16 had to be shifted by 7 ppm for CH_2, -1 ppm for CH_3 and 11 ppm CH to correct for errors in the absolute shifts. Although these errors may seem large they are still modest compared with the physically achievable range of the isotropic ^{13}C chemical shift of ≈ 200 ppm. The errors are largely due to the small atomic basis employed. This was established by performing time-consuming calculations with a larger atomic basis of approximate "triple zeta" quality (II', see Sect. 2.4.2) for a few selected geometries. In Fig. 31, the chemical shift obtained with this basis is plotted versus the standard "double zeta" chemical shifts for each residue CH_3, CH_2, CH separately. Evidently, the chemical shifts in the two bases are strongly correlated. For the simulation of geometry effects it is thus justified to replace the large, more realistic and very time-consuming basis II' by the small and much more convenient basis DZ. The defects of the small basis may be quantified by fitting the data to the function

$$\delta_X^{II'} = b_X \delta_X^{DZ} + a_X \qquad (4.2)$$

The results are shown in Table 7. The slopes b_X meet unity almost exactly for all chemical sites X, and only the offset a_X depends on the group. We may use these data to improve the DZ chemical shifts. If we impose unity slope ($b_X = 1$), the addition of the offset a_X substantially reduces the error in the absolute chemical shifts to -0.5 ppm for CH_3, and 2.5 ppm for CH and CH_2. The residual errors are expected to decrease further if even larger atomic basis sets are employed. Unfortunately, for these atomic bases, the computing time requirements are prohibitive.

Another source of uncertainty is the empirical force field, which is only the second-best way of geometry optimization. Again, the calculation on a

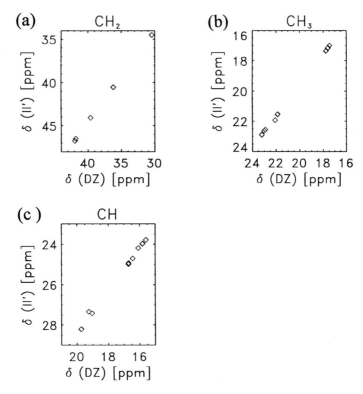

Fig. 31. The chemical shift obtained with the medium sized 3ζ-basis II' is plotted against the DZ chemical shifts for the various groups and for a few selected geometries of aPP. We find an excellent linear correlation with unity slope. The ordinate shifts, however, are non-zero and depend on the chemical group. This shows that the small basis DZ does not yield correct absolute shifts. Relative shifts are adequately calculated even with the small atomic basis

Table 7. The chemical shift with basis II' is correlated with the DZ shifts by the linear relation $\delta_X^{II'} = b_X \cdot \delta_X^{DZ} + a_X$. The ordinate shifts a_X depend on the type of the resonance, which reflects the smallness of the basis DZ. In all cases, the deviations from unity slope ($b_X = 1$) are negligible. If $b_X = 1$ is enforced (right part), then the ordinate shifts change. "dev" denotes the mean square deviation of the fit and the data points

X	b_X	a_X	dev.	a_X ($b_X = 1$)	dev.
CH_2	1.05	2.2	0.05	4.46	0.25
CH_3	1.02	−0.8	0.02	−0.35	0.07
CH	1.07	7.1	0.04	8.28	0.25

sufficient ab initio level, e.g. by SCF-MP, is still very costly in practice in spite of being desirable in principle. As a consequence, we must take refuge to the empirical force field method. Then, the reliability of the geometries depends directly on the quality of the parametrization. For this study, the widespread CVFF force field of Biosym Inc. (see Sect. 2.3) has been used. The impact of small changes in the parametrization is demonstrated in Fig. 32. There the chemical shift of geometries which have been optimized with zero atomic partial charges (no Coulomb interaction) is compared to the chemical shift of geometries obtained with a correct parametrization (including non-zero partial charges). The corresponding chemical shifts are strongly correlated, indicating that subtleties in the parametrization do not severely impede our approach. A comparison of the CVFF and AMBER force fields will be presented in Sect. 5.2.

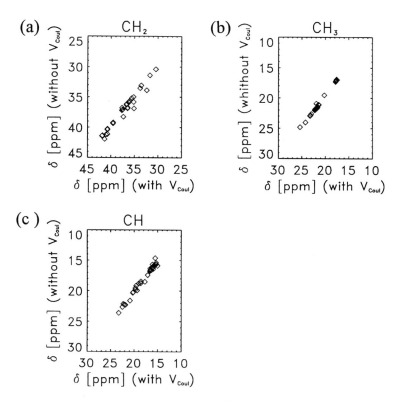

Fig. 32. The impact of variations in the force field parametrization (used for geometry optimization) is demonstrated. The chemical shift of geometries obtained without Coulomb interaction are well correlated with the chemical shift of the geometries optimized with the full parametrization. The observed scatter indicates that small effects might be obscured (or brought about) by a defective force-field parametrization

4.4
The Role of the Conformational Statistics

A simulated solid state spectrum basically displays the distribution of geometries that are encoded by the chemical shift. The conformational statistics discussed in this paper is only a first approximation of the physical geometry distribution function. Thus, the comparison of the simulated spectrum with experiments permits the check of the statistical model on the very short length scale of a few bond lengths. This is complementary to the scattering methods (X-ray, neutron scattering) which probe amorphous samples on a larger length scale (typically > 10 Å). Apart from shortcomings of geometry optimization and quantum chemistry (discussed in the previous sections of this chapter), the coincidence or non-coincidence of simulated spectrum and experiment is a *direct* way to assess the physical reality of a statistical model on a strictly local scale. The simulation presented in Fig. 16 is based on the 5-state RIS model of Suter and Flory [86] which describes a single chain without excluded-volume interaction of atoms separated by many segments of the chain [5]. Experimentally this description corresponds to a dilute polymer solution in a so-called ϑ-solvent where solvent effects and excluded-volume effects exactly cancel [14, 5]. The model [86] has been quite extensively checked, the characteristic ratios are well met although there seem to be some difficulties with temperature coefficients [87]. Fortunately, a single chain description is valid for the amorphous solid state, too. From a static point of view, solution, melt and glassy amorphous state are very similar and may be alike described by the random coil picture and RIS models. For large length scales (> 10 Å), the validity of this assertion has been established (within less than 20% deviation) by scattering methods for a large number of polymers [15, 18, 17], and especially for (isotactic) poly(propylene) [88]. The parameters obtained by these experiments (radius of gyration, characteristic ratio) mostly coincide within a few percent with the values obtained for the respective polymer solutions. The five state model is thus experimentally verified on large length scales. Up to now, there is no independent experimental method to obtain the local conformational statistics in the solid state. A bulk simulation by Theodorou and Suter [86], however, is in semiquantitative agreement with the RIS-model. The ratio of probabilities for the t- and g-states is the same as in the RIS model, although the $\bar{g}$-, t^*-, and g^*-states are more populated in the aforementioned simulation than in the RIS model. To summarize, the 5-state RIS model seems to be a reasonable description of the solution and the polymer glass both on a local and an extended length scale. These independent findings are strongly supported by the remarkable agreement between simulated and experimental solid state NMR spectra. Thus, together with the other experimental and theoretical techniques, the new method described in this paper may indeed be used to assess a statistical model of the geometry distribution.

The simulation was based on the 5-state RIS model mainly because it is, at the moment, the best statistical description of atactic poly(propylene). It should be borne in mind, however, that the ab initio simulation technique is not restricted to RIS models. Any model with sufficient predictive power may be used as a basis for the simulation: The conformational (or more generally: geometric) statistics may be replaced by the results of a molecular dynamics or Monte Carlo study, or any other method. Within computer time limitations, the relevant geometries may be fed into Eq. 3.15 and used as a basis for the quantum chemistry calculations.

4.5
Limiting Factors and Computer Time Considerations

The practical applicability of the method is limited both by physico-chemical facts and – perhaps even more severely – by computer time considerations. The latter naturally depend to a large degree on the available computing power. The following remarks refer to a modern workstation (SGI IRIS IN-DIGO, DEC ALPHA AXP).

1. The essential physical condition for a meaningful test of the simulation is an experimental spectrum that meets the assumptions of the simulation. The NMR lineshape should result from the inhomogeneous distribution of geometries and should not be distorted by experimental effects (motional broadening, insufficient proton decoupling, transversal relaxation, CP efficiency). Complete motional averaging (i.e. liquid-like behavior), however, can easily be taken into account.
2. The direct electronic packing effects should be small. In polar systems or samples with hydrogen bonding, problems might occur.
3. The necessary computer time for the IGLO method increases with v^3 to v^4, where v is the overall number of atomic basis functions. The size of the system that can be handled is thus strongly limited. The calculation of a single geometry for the model molecule (Fig. 15.c) with 19 carbon atoms and 40 hydrogen atoms took $\approx$ 12 h on a SGI IRIS INDIGO (MIPS R4000 CPU), and 4 h–6 h on the DEC ALPHA AXP 3000 series (models 400, 600, 800; 175/200 MHz). A calculation with the larger basis II' took $\approx$ 90h on SGI IRIS INDIGO.
4. On a workstation or a small workstation cluster, the number of relevant geometries should not exceed a few hundred. Quasicontinuous distributions imply the calculation of a larger number of geometries and may be intractable today except for the most simple systems.

5. The IGLO method is a Hartree-Fock technique that neglects correlation effects. If these effects are appreciable, the standard IGLO method cannot be used and must be replaced by either GIAO-MBPT (the Møller-Plesset version of GIAO) [59, 60] or MC-SCF IGLO [40, 42]. Unfortunately, there is no general rule to predict the strength of correlation effects; double bonds should be regarded with special suspicion. In some systems with heavy atoms, relativistic effects are imminent [40].

6. As long as the geometries are generated by a force field minimization, it has to be checked whether the various bonding situations within the molecule are properly parametrized. For uncommon or unclear bonds (partial conjugation, systems with heavy or rarely used atoms), the force-field parametrizations are very often defective. In these cases at least, a suitable ab initio optimization should be performed.

7. In view of the various approximations and the sources of error, small effects (deviation of the chemical shift < 1 ppm, variations in intensity < 10%) cannot be properly investigated.

For the simulation of amorphous polymers in the solid state, these limitations are not very restrictive, as most commercially important polymers display a simple bonding situation that is well parametrized by modern force fields. They contain mostly atoms of the first two rows of the periodic system and can be mostly treated by non-relativistic Hartree-Fock level calculations. It is emphasized, however, that the above requirements should be checked for any system in advance.

4.6
Simulation of Other Amorphous Systems

Apart from polymers, many classes of compounds display amorphous or glassy phases (mineral oxidic glasses, metallic glasses). At least for mineral glasses, the observed disorder is molecular in origin. For example in silicate or phosphate glasses, there are a number of structural units, the so-called Q-units that differ in the number of oxygen bridges. Moreover, large deviations of the bond lengths and the bond angles from the ideal values are found. In the NMR spectra of the spin-$\frac{1}{2}$ nuclei (^{29}Si, ^{31}P), this inhomogeneity shows up as a broad and partially structured spectrum. For an introduction to the topic and a guide to the original literature we refer to the recent review by Eckert [99]. The different structural units can be distinguished, sometimes even the kind of the counter ion (in binary or ternary glasses). The various peaks themselves display an inhomogeneous broadening of typically 10 ppm which is not connected to the binding situation (Q-unit, counter ion) but is mostly attributed to variations in geometry. Recent theoretical investigations [78, 77, 70] support this view.

The situation in mineral glasses is thus comparable to amorphous polymers, as, in both sytems, molecular disorder and variations in geometry lead to an inhomogeneous broadening in the solid state spectrum. A strategy analoguous to the technique described in this study should be able to elucidate the local structure in these mineral glass systems – at least in principle. In practice, some obstacles have to be overcome. First of all, a precise geometrical model is needed as a basis for a simulation. Most likely, such a model will dwell on continuous variations of the relevant parameters (bond lengths, bond angles) which increases the number of relevant geometries.

But the most severe hindrance for a simulation is the fact that mineral glasses form a *three-dimensional* network whereas ordinary synthetic polymers are *one-dimensional* chains. In a three-dimensional network both the size of a representative segment and the number of relevant geometries will be tremendously larger than in a chain molecule. The necessary computer time will be accordingly large.

Moreover, two aspects of a more technical kind must be regarded. Mineral glasses contain relatively heavy elements that are difficult to tackle by ordinary force fields and by standard quantum chemistry methods. This may enforce the use of time-consuming advanced methods.

Another problem is posed by the shape of the experimental resonances. They are normally almost perfectly Gaussian and bear very little structure to discriminate competing geometrical models.

To conclude, the ab initio simulation of NMR spectra is much harder in mineral glasses than in polymer systems. For the latter, geometrical models can already be simulated with present day computing power and by a proper combination of established methods. For the former, it is much more convenient to investigate the influence of varying bond lengths and widened bond angles by parameter variation without embarking on a full scale simulation of a large mineral glass segment.

In this section, selected results for other polymers will be presented. It will be demonstrated that our method is applicable to a wide range of systems belonging to different structural classes, including the simplest polymer poly(ethylene) (PE), a bisubstituted polyolefine (poly(isobutylene), PIB), unsaturated systems (poly(butadiene), PBD and poly(isoprene), PIP), polymers with hetero atoms (poly(vinyl chloride), PVC), and extended sidegroups (poly(methyl methacrylate), PMMA).

All compounds (including aPP presented in the previous chapter) were used as purchased. For the polymers with a glass transition temperature T_g below room temperature (aPP, PIB, PBD, PIP, PE), the sample was slowly cooled down in the course of the NMR measurement, and the spectra were recorded after the glass transition temperature was reached. If T_g is above room temperature (PVC, PMMA), the samples were measured without prior treatment at ambient temperature.

If technically possible, all samples were heated and cooled down again. In neither case, the amorphous solid state spectrum changed appreciably after performing these mild temperature programs. No special procedures (like precipitation from solution) were chosen to influence the microstructure of the glassy amorphous state.

5.1
Poly(ethylene)

Poly(ethylene) (PE) (Fig. 7) is the simplest polymer and permits the investigation of fundamental questions. PE is a semicrystalline compound, and in a ^{13}C solid state NMR spectrum normally both the crystalline phase (around 33 ppm) and an amorphous contribution ($\approx$ 31 ppm) can be observed (Fig. 33.a). At room temperature, the amorphous resonance is motionally averaged. The glass transition temperature in this semicrystalline system is still a matter of some controversy [100, 101, 102], but most likely the calorimetric glass point is somewhat below 200 K, perhaps as low as 150 K. Experimentally, the recording of the ^{13}C solid state NMR spectrum of the amorphous

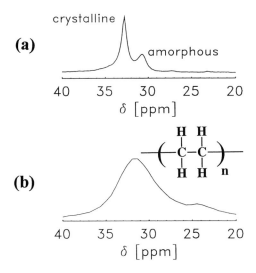

Fig. 33. ^{13}C-CP/MAS NMR spectrum of poly(ethylene) at room temperature (**a**). Contributions of both the crystalline phase and the amorphous phase are discernible. At room temperature, the amorphous phase is still motionally averaged. To obtain a spectrum of glassy amorphous poly(ethylene) at $T \approx 200$ K, a T_1-filter has to be applied which suppresses the crystalline contribution (**b**). At this temperature, the spectrum is most likely inhomogeneously broadened (see text)

glassy contribution requires special techniques, as, at T = 200 K, the standard CP/MAS experiment strongly favors the crystalline contribution over the amorphous signal. The latter may be enhanced, however, by applying a T_1-filter making use of the much longer longitudinal relaxation times in the PE crystallites ($T_1 > 100$ sec compared with a few seconds in amorphous PE). The spectrum displayed in Fig. 33.b has been obtained by this technique on a BRUKER ASX 500 at $T \approx 200$ K and with a repetition time of 200 msec.

The spectrum consists of a single, rather featureless Gaussian-shaped peak with a width of about 5 ppm (at half height); the small shoulder at ≈ 25 ppm represents mobile (methyl) endgroups. In the following we assume that the observed broadening is due to the inhomogeneous superposition of the chemical shifts belonging to different geometries. Experimental shortcomings like insufficient proton decoupling or motional narrowing should be relevant only at higher temperatures [103, 104], and the recorded spectrum should be essentially unaffected by these artifacts.

In a way, poly(ethylene) is *the* typical chain molecule and has attracted a lot of theoretical interest. Fortunately, the various model descriptions of the conformational statistics are quite similar [105, 5, 27]. The models assume three rotational isomeric states (t, g^+, g^-) with slightly different values for the

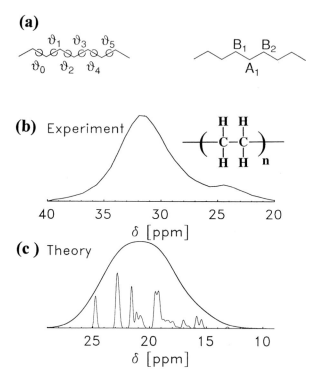

Fig. 34. Model molecule for PE, definition of the dihedral angles and label of the atoms in the central part of the model molecule (**a**). Experimental (**b**) and simulated (**c**) solid state spectrum of poly(ethylene). The *fine lines* reflect the contributions of the various conformations (Eq. (3.15), slightly broadened for typographical reasons), whereas the *bold line* is the overall simulation (Eq. (3.16)). The width of the resonance is slightly overestimated by the simulation. The small upfield peak in the experimental spectrum (at ≈ 24 ppm) is most likely due to endgroups which are much enhanced by the technique of recording

gauche states ($\pm 112.5°$ [105], $\pm 120°$ [5], $\pm 115°$ [27]); the *trans* state is always set to $\vartheta = 0°$. The statistical models are in good agreement with experimental results (neutron scattering) [27, 106, 15] as well as with bulk molecular dynamics simulations of PE melts [107, 108]. In this study, we have adopted the most recent model [27] as a basis for this simulation. In [27], the conformational statistics is encoded in a matrix of *a priori* probabilities p_{ij}. The probabilities for longer sequences are calculated via Eq. (2.14). Unfortunately, only a matrix for $T = 300$ K is given which introduces some error in the conformational statistics when describing the glassy polymer at $T = 200$ K. The model molecule for the simulation is depicted in Fig. 34.a; its geometry is fully determined by a sequence of 6 dihedral angles ($\vartheta_0, \vartheta_1, \vartheta_2, \vartheta_3, \vartheta_4, \vartheta_5$), all of which are controlled in the course of the simulation. Again, only the central atoms

A, B_1, B_2 are in a sufficiently "polymer-like" position to be included in the simulation. The quantum chemical computations are facilitated by the symmetry elements of the molecule. Neither the inversion of the "read"-direction

$$(\vartheta_0, \vartheta_1, \vartheta_2, \vartheta_3, \vartheta_4, \vartheta_5) \longrightarrow (\vartheta_5, \vartheta_4, \vartheta_3, \vartheta_2, \vartheta_1, \vartheta_0) \tag{5.1}$$

nor a reversed sense of rotation of the dihedral angles

$$\vartheta_i \longleftrightarrow -\vartheta_i, \qquad\qquad t \leftrightarrow t, \qquad g^+ \leftrightarrow g^- \tag{5.2}$$

affect the chemical shift of the central carbon A. Using this symmetry, about 60 different conformations need to be considered. As for aPP, the geometries are obtained by a force-field minimization with the empirical CVFF force field. For the quantum chemistry calculations, the basis set DZ was used. Because of the small number of atomic basis functions, only roughly 1 hour of CPU time (SGI IRIS INDIGO) was necessary for every geometry. The simulated spectrum was calculated via Eq. (3.16). The width of the Gaussian broadening function was 1.5 ppm. In Fig. 34.b/c, the experimental spectrum is compared with the simulation. It can be seen that the simulation is somewhat broader and has a slightly different shape. Possible reasons for the deviations are distortions of the experimental spectrum by the T_1-filter, residual motional averaging, and the use of a statistics for $T = 300$ K in place of $T = 200$ K. The rather unspectacular shape of both experimental and theoretical spectrum precludes any discussions on the validity of the underlying statistical model.

More importantly, in the simple system poly(ethylene), the influence of the geometry can be thoroughly investigated. As a first point, the γ-*gauche* effect can be observed (Fig. 35.a), i.e. the chemical shift of the central atom A is correlated with $|\vartheta_1| + |\vartheta_4|$. As in the case of aPP, for fixed $|\vartheta_1| + |\vartheta_4|$ an appreciable spread of ≈ 4 ppm in the chemical shift is found, which is indicative of the importance of other geometric parameters. In PE, the source of this spread can be identified. In Figs. 35.b-d, the chemical shift is plotted vs. the internal dihedral angles $|\vartheta_2| + |\vartheta_3|$ for each of the clusters separately. A very good correlation is observed with a very small residual jitter. This finding demonstrates that next to the traditional γ-*gauche* sensitive dihedral angles ϑ_1 and ϑ_4, the interior angles ϑ_2 and ϑ_3 are important for the chemical shift whereas the influence of the remote dihedral angles ϑ_0 and ϑ_5 is minute. It should be noted that ϑ_2 and ϑ_3 are quite important for all of the systems investigated, but only in PE a simple correlation of these parameters and the chemical shift could be established.

As the various geometrical parameters are strongly interrelated, additional secondary correlations of other geometrical parameters with the chemical shift emerge. δ is apparently correlated with the added lengths of the four

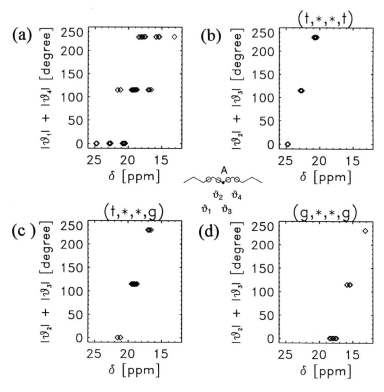

Fig. 35. Poly(ethylene) displays a γ-*gauche* effect, as the chemical shift of the central methylene carbon A is correlated with $|\vartheta_1| + |\vartheta_4|$ (**a**). The horizontal clusters betray the importance of other geometric parameters. In PE, these can be shown to be the interior dihedral angles ϑ_2, ϑ_3 as the chemical shift of each of the clusters is correlated with $|\vartheta_2| + |\vartheta_3|$ separately (**b, c, d**)

adjacent bonds (Fig. 36.a), the added distances of the carbon β-neighbors (Fig. 36.b), the γ-neighbor distance (Fig. 36.c), and the two adjacent bond angles (Fig. 36.d).

As the treatment of a single geometry is computationally not very expensive, maps of the chemical shift as a function of the dihedral angles may be calculated (Fig. 37). For each of these maps, the external dihedrals ϑ_1 and ϑ_4 were varied independently from 0° to 360° in 15° steps; the maps are thus based on a grid of $24 \times 24 = 576$ geometries. The internal angles (ϑ_2, ϑ_3) were set to (t, t) (top), (t, g^+) (middle), (g^-, g^+) (bottom), respectively. Prior to the IGLO calculations the geometries were again optimized with the CVFF force field, and the dihedral angles were constrained to their grid values. In these calculations, symmetry relations were *not* employed to reduce the number of relevant geometries, and can be used as a consistency check of the results.

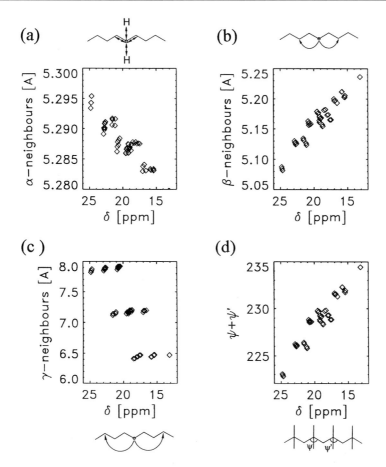

Fig. 36. The chemical shift of the methylene unit in PE is correlated with the combined bond lengths of the adjacent bonds (**a**), the distance of the carbon β-neighbors (**b**), the distance of the γ-neighbors (**c**) and the sum of the adjacent bond angles $\psi + \psi'$ (**d**). These numerous correlations occur because the various geometry parameters are mutually interrelated. The clear correlations reflect the simple geometric and steric situation in PE

On examining the maps, the following observations can be made:

- Both the chemical shift map and the energy map reflect the symmetry of the molecule as can be seen especially for $(\vartheta_2, \vartheta_3) = (t, t)$.
- *trans* states are less sensitive to small variations of the dihedral angle than *gauche* states. In the simulations of the NMR spectra, a unique Gaussian broadening function was used for all conformational states. By this assumption, the number of parameters was kept small, but in the maps it can be seen that this assumption is in fact a simplification.
- In some respects, the chemical shift maps are similar to the energy maps.

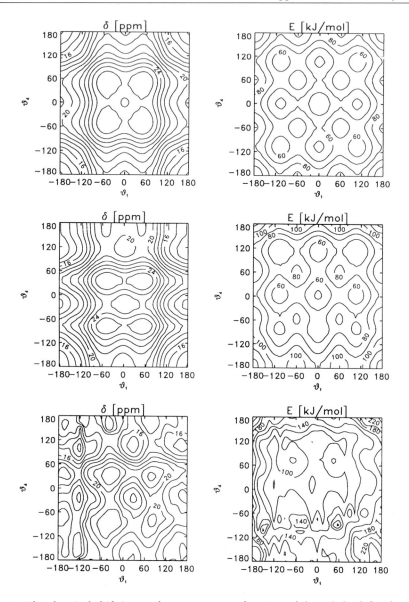

Fig. 37. The chemical shift in PE (for atom A) as a function of ϑ_1 and ϑ_4 (*left column*). The interior dihedral angles have been set to $\vartheta_2 = t$, $\vartheta_3 = t$ (*top row*), $\vartheta_2 = t$, $\vartheta_3 = g^+$ (*middle row*), and $\vartheta_2 = g^-$, $\vartheta_3 = g^+$ (*bottom row*), respectively, and have been kept fixed. The maps have been calculated by systematically varying ϑ_1, ϑ_4 on a grid of $15° \times 15°$. The maps belonging to different interior angles differ appreciably. The apparent symmetry in some of the maps (especially for ($\vartheta_2 = t$, $\vartheta_3 = t$)) is a result of the calculation and has not been imposed. The symmetry of the other maps (with *gauche* interior dihedral angles) is considerably reduced.

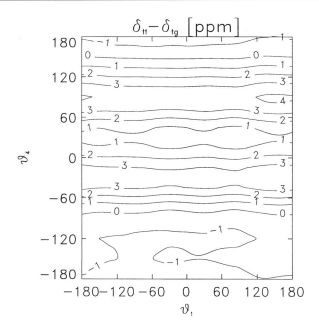

Fig. 38. Map of the difference $\delta(\vartheta_1, t, t, \vartheta_4)$ - $\delta(\vartheta_1, t, g^+, \vartheta_4)$ for PE. It is seen that the difference is not just a constant, but itself a function. A two dimensional chemical shift map as a function of ϑ_1 and ϑ_4 is thus of little use as the influence of the interior dihedral angles is deliberately neglected

- From the energy maps, it can be seen that the largest part of the $(\vartheta_1, \vartheta_4)$-plane is energetically inaccessible at normal temperatures. The chemical shift values calculated for these areas are thus of little practical interest.
- In spite of a superficial similarity among the maps in Fig. 37, the detailed structure of the maps depends on the dihedral angles ϑ_2 and ϑ_3 as well. Changing these dihedrals from *trans* (top) to *gauche* (bottom) introduces new extrema and effects a general distortion of the chemical shielding surface. The contribution of the internal dihedral angles is not a simple constant as can be seen from the difference $\delta(\vartheta_1, t, t, \vartheta_4) - \delta(\vartheta_1, t, g^+, \vartheta_4)$ (Fig. 38, calculated from the maps in Fig. 37). As a consequence, the two-

---

It is seen that the chemical shift changes steeply, if either ϑ_1 and ϑ_4 is in a *gauche* state whereas the variation around the $\vartheta_1 = t$, $\vartheta_2 = t$ state in the center of the map is only smooth.

The map of the corresponding forcefield energies are shown for comparison (*right column*), the symmetry is the same as in the chemical shift maps. It can be seen that most of the $(\vartheta_1, \vartheta_4)$-plane is unattainable at thermal energies. The energy map for $(\vartheta_2 = g^-, \vartheta_3 = g^+)$ (*bottom*) suffers from shortcomings of the contour line algorithm

dimensional maps still do not give the full information about the geometry dependence of the chemical shift. The full function $\delta(\vartheta_1, \vartheta_2, \vartheta_3, \vartheta_4)$ on a $15°$-grid, however, requires $24^4 = 331776$ geometries which would take several years on a currently available workstation.

From the last remark it becomes clear that consistent and complete chemical shift maps for amorphous polymers are computationally very costly, even more so for more complicated systems. And even if such a map existed, it must be kept in mind that a fair number of the geometries are fictitious as their high energy implies a very low Boltzmann factor. Moreover, the reliability of force-field geometries is questionable for high-energy states. Thus chemical shift maps pose a lot of questions whereas their practical advantage is limited. The RIS states really occupied in nature can be assessed with a much lower expenditure, for what reason we have chosen the pragmatic approach described in chapter 3. In polypeptides and biomolecules, the situation is different, the occurring difficulties are somewhat relaxed, and chemical shift maps may be calculated with some success [73].

5.2
Poly(isobutylene)

5.2.1
Simulation and Comparison with Experiment

Poly(isobutylene) (PIB) is a bisubstituted polyolefine and has a very crowded structure. Under normal circumstances, it does not crystallize and undergoes a glass transition at $T_g \approx 208$ K. [85] The experimental CP/MAS spectrum of glassy amorphous PIB (Fig. 39) displays a large and partially structured broadening of the methylene and methyl position. A dynamic origin of this lineshape can be ruled out experimentally by 2D exchange experiments [63]. Thus, the observed spread reflects the geometrical disorder in the amorphous phase.

The two methyl sidegroups imply a large steric hindrance and correspondingly complicated conformational statistics. Early attempts to devise an RIS model [109, 23, 110] started from a systematic analysis of the energy hypersurface and provided unsatisfactory results. Only by utilizing a Monte Carlo sampling technique [26] could a better agreement with experiment (especially neutron scattering curves) be achieved. The model of Vacatello and Yoon considers 6 rotational isomeric states (Table 8), as each of the traditional states t, g^+, g^- splits into two due to the large steric interactions.

Experimentally, as in aPP, the methylene and methyl region show the largest effect. Therefore, we have chosen the "tetramer" depicted in the insert of Fig. 40 as model molecule for the extended polymer chain. The γ-neighbors

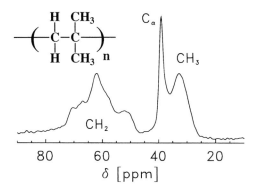

Fig. 39. Experimental CP/MAS spectrum of poly(isobutylene) recorded at $T = 200$ K, below the calorimetric glass transition temperature. A considerable inhomogeneous broadening is experienced. The methylene region is richly structured. The quarternary carbon resonance is quite narrow and will not be discussed

Table 8. Rotational isomeric states assumed in the model [26] for poly(isobutylene). The model is based on a continuous Monte Carlo simulation, and the various states are the centers of regions of elevated probability

Name	t^+	t^-	g_+^+	g_-^+	g_+^-	g_-^-
ϑ	$+15°$	$-15°$	$+130°$	$+105°$	$-105°$	$-130°$

of the said groups are well included within the model molecule. The geometry of the model molecule is almost completely determined by specifying the dihedral angles $(\vartheta_1, \ldots, \vartheta_4)$, as the possible low energy states of ϑ_0 and ϑ_5 are nearly equivalent because of the high symmetry of the terminal isobutyl groups.

In [26], the conformational statistics is encoded by a pair of *a priori* matrices, p'_{ij} and p''_{ij} (In PIB, two matrices are necessary as there are two different types of main chain atoms, methylene and quaternary). As for PE, the probability for a larger sequence of dihedral angles can be calculated by a modified version of Eq. (2.14).

$$p(\Theta_{i_1}, \ldots, \Theta_{i_n}) = p_{i_1} q''_{i_1, i_2} q'_{i_2, i_3} q''_{i_3, i_4} \cdots q'_{i_{n-1}, i_n} \tag{5.3}$$

In this formula, p denotes the vector of *a priori* probabilities for a single dihedral angle, and q'_{ij}, q''_{ij} the respective matrices of conditional probabilities. In the numerical implementation, Eq. (5.3) was taken as a basis for the generation of a large number of chains by a random sampling technique, and $p(\Theta_{i_1}, \ldots, \Theta_{i_n})$ was obtained as a numerical average.

Unfortunately, the *a priori* matrices were available only for $T = 400$ K which differs markedly from the experimental glass transition temperature

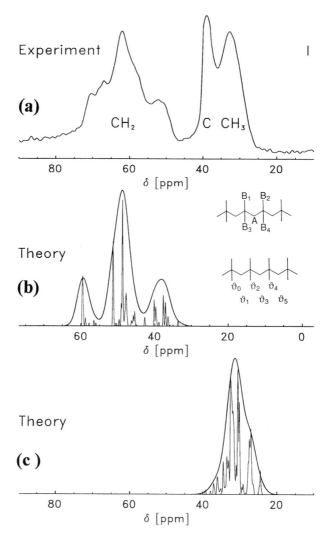

Fig. 40. Experimental (**a**) and simulated (**b, c**) spectrum of PIB. For the simulation the six state conformational statistics of Vacatello and Yoon [26] for $T = 400$ K has been used, whereas the experimental spectrum was recorded at $T = 200$ K. The *fine lines* reflect the contributions of the various conformations (Eq. (3.15), slightly broadened for typographical reasons), whereas the *bold line* is the overall simulation (Eq. (3.16)). The width of the methylene and methyl region is correctly reproduced as well as the occurrence of three main subpeaks in the methylene region. The region between the subresonances is underestimated which might be a consequence of the elevated temperature of the conformational statistics. The methine region is narrow experimentally and is not investigated theoretically.

The insert displays the model molecule for PIB with the labels of the atoms in the center of the molecule. They are considered representative of an extended chain. The relevant dihedral angles are depicted

$T_g = 208$ K. This is why a full quantitative agreement of experiment and simulation cannot be expected. The calculations reveal that ≈ 150 different conformations contribute with a probability of > 0.001 each; the overall probability of these geometries is > 0.92. The number of geometries to be treated explicitly by quantum chemistry is considerably reduced by the symmetry of the molecule, i.e.

- reversing the "read" direction

$$(\vartheta_1, \vartheta_2, \vartheta_3, \vartheta_4) \longrightarrow (\vartheta_4, \vartheta_3, \vartheta_2, \vartheta_1) \tag{5.4}$$

- changing the sense of rotation of the dihedral angles

$$\vartheta_i \longleftrightarrow -\vartheta_i, \qquad t^+ \leftrightarrow t^-, \qquad g_+^+ \leftrightarrow g_-^-, \qquad g_+^- \leftrightarrow g_-^+ \tag{5.5}$$

Then, 49 conformations are left for an explicit treatment. The corresponding geometries were obtained by a force field minimization (CVFF) with constrained dihedral angles $\vartheta_1, \ldots, \vartheta_4$. The quantum chemical calculations (IGLO) were performed in the DZ basis (242 basis functions). One geometry took about 12 hours on a SGI IRIS INDIGO. The simulated spectra were obtained by Eq. (3.16); the width of the Gaussian broadening function was 1.5 ppm for all sites [9].

The simulated spectrum is in good qualitative agreement with experiment (Fig. 40). The overall width of the methylene and methyl positions, respectively, are well reproduced. The splitting of the methylene region into three main contributions is present both in the experimental and the theoretical spectrum. The calculation is unable to reproduce the finer details of the experiment, especially the intensity in between the main subresonances is underestimated. The deviations can probably be tracked back to the high temperature of the underlying statistics. Naturally, shortcomings of the model as a whole will also affect the simulated spectrum. Further investigations will be needed to clarify this point [111]. The narrow quaternary carbon signal is not extensively discussed as its conformational environment is not fully specified by $\vartheta_1, \ldots, \vartheta_4$. The γ-neighbors of the quaternary carbon atoms are, moreover, in an atypical terminal position in the model molecule. An inspection of the quaternary chemical shift reveals a very small spread (< 2 ppm), again in good agreement with experiment. As in the other systems, the absolute chemical shifts are not properly reproduced. Again this is mainly due to the small atomic basis employed.

The ab initio method of spectral simulation allows the discrimination of different statistical models. For PIB, a competing RIS model [23, 110] assumes only four states (t^+, t^-, g_+^+, g_-^-) and a simplified statistics. In Fig. 41, a simulation based on this different model, but the same set of geometries

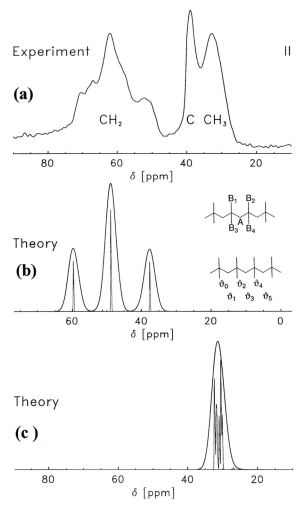

Fig. 41. Experimental (**a**) and simulated (**b, c**) spectrum of PIB. For the simulation the four state RIS model of Suter, Saiz and Flory [23] has been used. Compared with Fig. 40, the overall agreement of experiment and theory is much worse. The width of the resonances is underestimated and there is no appreciable intrinsic broadening due to different conformations

and chemical shift calculations, is compared with experiment. The qualitative agreement is considerably worse than in Fig. 40; especially the width of the CH$_3$-resonance is grossly underestimated. Although the basic splitting of the methylene resonance is reproduced, there is only very little intensity in between the subresonances. Apart from the Gaussian convolution function (introduced "by hand"), there is no intrinsic broadening due to the statistical model. Thus, the overall agreement with experimental spectra is worse in the four-state model compared with the six-state model. In spite of some minor shortcomings, the latter appears to be a more accurate description of the conformational statistics in PIB.

5.2.2
Geometry Effects

The methylene group (A in Fig. 40 (insert)) displays a *y-gauche* effect (Fig. 42); the chemical shift is correlated with the dihedral angles ϑ_1 and ϑ_4. The six horizontal clusters of geometries with approximately the same value of $| \vartheta_1 | + | \vartheta_4 |$ stem from the combination of two *trans* states ($\pm 15°$) and four *gauche* states ($\pm 105°$, $\pm 130°$). The geometries underlying the sub-resonances can be identified. As in aPP, the downfield peak of the methylene region represents $(\vartheta_1, \vartheta_2, \vartheta_3, \vartheta_4) = (trans, *, *, trans)$ states, the central $(trans, *, *, gauche)$, and the upfield subresonance $(gauche, *, *, gauche)$; the unspecified dihedral angles ϑ_2 and ϑ_3 are denoted by an asterisk "$*$". Different from aPP, this assignment is based entirely on quantum chemical calculations, as there are no crystalline modifications of PIB for comparison. The

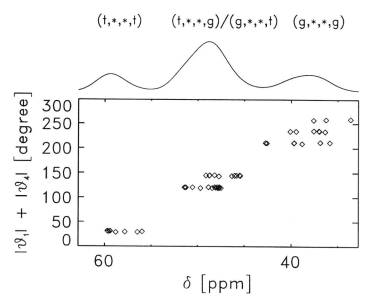

Fig. 42. The methylene resonance of poly(isobutylene) displays a *y-gauche* effect, i.e. a correlation of δ with $| \vartheta_1 | + | \vartheta_4 |$. The spread within the horizontal clusters is again indicative of effects which go beyond the traditional *y-gauche* effect. By identifying the geometries belonging to a given value of the chemical shift, it is possible to perform a subpeak assignment (*top of the plot*)

considerable spread within the said horizontal groups reveals the importance of geometrical parameters other than the traditional γ-*gauche* dihedral angles ϑ_1 and ϑ_4. Among these, the interior dihedrals ϑ_2 and ϑ_3 are the most relevant, as was demonstrated in an independent series of calculations. The exterior dihedrals were kept fixed and $(\vartheta_2, \vartheta_3)$ was varied systematically. As usual, the geometries were optimized with the empirical CVFF force field. The subsequent IGLO calculations revealed a complex dependence of the chemical shift δ on $(\vartheta_2, \vartheta_3)$ with an overall variation of ≈ 8 ppm, which is approximately the spread within the horizontal groups in Fig. 42. The interior dihedral angles are in fact very important, although their influence on the chemical shift cannot be parametrized by a simple function as in PE (Fig. 35).

The γ-partners of the methyl groups (B_i in Fig. 40 (insert)) are the quaternary carbons. According to the phenomenological γ-*gauche* effect, the chemical shift should be determined by the dihedral angles $(\vartheta_1, \vartheta_2)$ for B_1/B_3 and $(\vartheta_3, \vartheta_4)$ for B_2/B_4. The explicit calculation does not show any indication for such a correlation (Fig. 43.a). The reason for this failure of the phenomenological concept in the "crowded" polymer PIB can be traced back to the fact that the bond angles strongly depend on the adjacent dihedral angles and deviate from the ideal tetrahedral coordination (Fig. 43.b). The silent assumption underlying the phenomenogical γ-*gauche* effect is thus violated and the geometry cannot be described properly by a set of dihedral angles with all other parameters being kept to their textbook values. The distance of the methyl group to the γ-neighbors does not depend exclusively on the dihedral angles but also on the bond angles and (to a lesser degree) on the bond lengths. It can be seen that the geometry-dependent part of the chemical shift has a strong steric component for both the methylene and methyl carbons (Fig. 44). For the methylene group, this distance can well be represented by $| \vartheta_1 | + | \vartheta_4 |$ and by this mechanism, the traditional concept of the γ-*gauche* effect depending on dihedral angles rather than intramolecular distances is recovered. For the methyl carbons, this does not work due the strong steric interaction and the widening of the bond angles to values well beyond the ideal tetrahedral coordination.

To avoid confusion, it should be noted that the dihedral angles are still the essential parameter determining a geometry but the other geometric degrees of freedom (bond angles, bond lengths) depend on the dihedrals and strongly influence the *shape* of the molecule. Thus, the dihedrals may still be used to denote a certain geometric state by a unique set of numbers, but for the actual shape of the molecule in space (and the chemical shift which is a function of the latter), bond angles and bond lengths must be taken into account as well.

As the various geometric parameters are interrelated, unexpected correlations with the chemical shift can emerge. In PIB, the chemical shifts of the

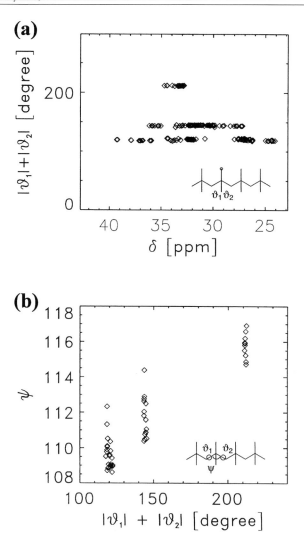

Fig. 43. The methyl carbon in PIB does not display a γ-*gauche* effect, as there is no apparent correlation of δ with $| \vartheta_1 | + | \vartheta_2 |$ (**a**). This failure of the phenomenological understanding of the geometry dependence of the chemical shift is due to the large steric interactions in this crowded system. The bond angle ψ strongly deviates from the ideal tetrahedral value and depends on $| \vartheta_1 | + | \vartheta_2 |$ (**b**). The influence of the dihedral angles is thus overcompensated

methyl and methylene carbons are both very well correlated with the lengths of the adjacent carbon bonds (Fig. 45). In a certain sense, this correlation is "accidential", as it is not the bond lengths themselves that *cause* a certain chemical shift, but the bond lengths are interrelated with other, more essen-

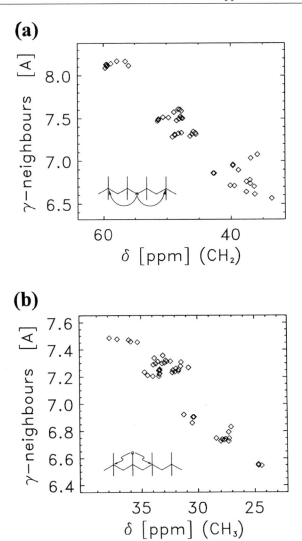

Fig. 44. The geometry dependence of the chemical shift is largely due to steric interactions as can be inferred from the correlation of δ with the γ-neighbor distance for the methylene (**a**) and methyl (**b**) resonance. The observed scatter betrays the presence of non-steric contributions to the chemical shift

tial degrees of freedom in such a way that eventually a very good correlation with δ emerges. The assertion that the bond length is not the *causal* reason for the chemical shift can be inferred from Fig. 46. Here, two geometries with a very high and very low value of δ, respectively, were selected, and the bond lengths d_1 and d_2 were varied with the other geometry parameters

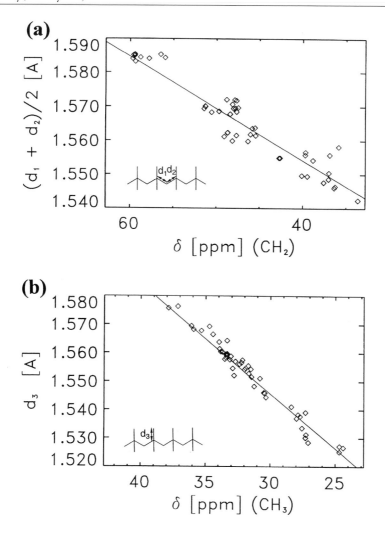

Fig. 45. Calculated chemical shift vs. the bond lengths of the adjacent carbon-carbon bonds for the methylene (**a**) and methyl (**b**) region. The set of optimized geometries computed for the spectral simulation was used for this plot. A strong correlation is observed. The *solid lines* are only guides to the eye and are not meant to suggest fits

unchanged. The chemical shifts of these artificial geometries are symbolized by ×, △, □, ◇, whereas the results of Fig. 45, which have been inserted for convenience, are denoted by ■. It is seen that by varying the bond lengths alone (×, △, □, ◇), a variation of the chemical shift of only about 5 ppm can be obtained. This must be compared with more than 20 ppm for the spread found in the geometries, where the variation of the bond lengths resulted

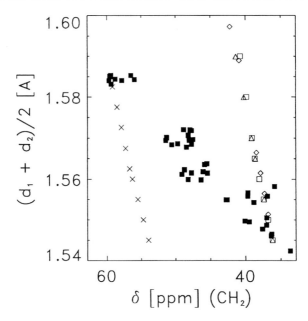

Fig. 46. The chemical shift of the methylene region is correlated with the length of the adjacent bonds of the optimized geometries (■, reprinted from Fig. 45). If a special geometry is selected, and only $(d_1 + d_2)/2$ is varied with the rest of the geometry unchanged ($\times, \square, \diamond, \triangle$ for various geometries), then the observed change of the chemical shift is much smaller although the spread in the bond length is the same as in the optimized geometries. This demonstrates that the original finding (■) is simply a *correlation*, and not a *causal explanation*. It is not the change of the bond lengths that brings about the variation of the chemical shift, but the bond lengths happen to be correlated with the more important geometric parameters (mainly dihedral angles) in such a way, that effectively a very neat correlation emerges

from a variation of the dihedrals followed by the optimization the geometry (■). This clearly shows that in Fig. 45, a *correlation* is depicted and not an *explanation*.

When considering this effect, a word of caution is in place. The geometries for this study were obtained by a constrained minimization in the CVFF empirical force field. Although this force field is a sophisticated, second-order parametrization which should provide reliable geometries, it could be defective in some respects. In the future (with a higher computing power becoming available), this problem should be reconsidered with more reliable ab initio geometries. Only then can the physical pattern be disentangled from force-field peculiarities.

5.2.3
Effects of the Atomic Basis, Endgroups, and Empirical Force Field

As in the other systems, the absolute value of the chemical shift is not well reproduced by the quantum chemical calculation. This is an artifact of the small atomic basis as can be seen from Fig. 47, where the same set of geometries has been recalculated with different atomic bases (II, II', DZ, and another *double zeta* basis DZ'). It is seen, that the geometry-induced variation is well reproduced by the small atomic basis sets whereas the absolute values change with the basis.

To check the influence of the endgroups on the chemical shift in the central part of the model molecule (A, B_i), an optimized geometry was taken, and one methyl group, two methyl groups, or two isobutyl groups were added to the molecule. If the new extended geometries are not optimized, the chemical shift in the central part is virtually unchanged (variation < 0.2 ppm). The direct electronic influence of these atoms is thus negligible and no appreciable error is introduced by using the model molecule as a substitute for the macromolecule. If the extended geometries (with the additional methyl and isobutyl groups) were reoptimized, the deviations were larger, but still quite small (< 1 ppm). This is the *indirect* influence of the endgroups which works by changing the geometry in the central part. It is still smaller than

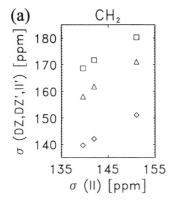
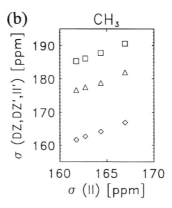

Fig. 47. The chemical shifts obtained with smaller and less time consuming basis sets ($\Diamond$:II', $\triangle$: DZ' (a special basis of 2ζ quality), $\square$: DZ (the ordinary basis used throughout this review)) are well correlated with the chemical shifts obtained with the large basis II. As the correlation is linear with almost unity slope, the small basis can be used for the calculation of relative geometry effects. The absolute values are not properly reproduced by small basis sets as can be seen from the ordinate shifts that depend on the residue involved. Note, that in this plot, the chemical shift is given on the theoretical scale σ rather than on the experimental scale δ

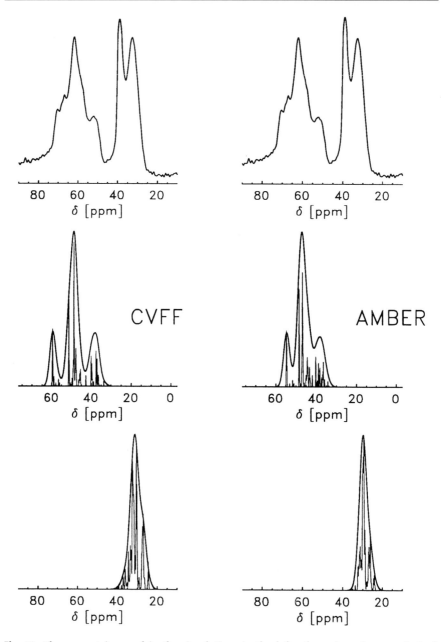

Fig. 48. The geometries used in the simulations in the *left column* have been optimized with the CVFF force field whereas the *right column* relies on AMBER geometries. The overall agreement with experiment is comparable. The simulated AMBER spectra are somewhat smaller than their CVFF counterparts

the broadening function used in the spectral simulation and does not impair the results.

The impact of the forcefield parametrization on the results can be assessed by using two different empirical force fields for the geometry optimization, the rest of the procedure remaining unchanged. As a suitable candidate for comparison, the AMBER force field was chosen which has found a widespread use in molecular dynamics simulations. Compared to CVFF, it lacks cross terms that couple bond lengths, bond angles and dihedral angles by explicit energy functions (see Sect. 2.3). The quality of the resulting geometries should be somewhat poorer with AMBER due to the simplistic parametrization.

Qualitatively, the two force fields yield the same results as can be seen from the comparison of the simulated spectra (Fig. 48). The qualitative results, upon which we put the greatest emphasis in the present study, are unaffected by the choice of the force field. A closer look, however, reveals that there is no 1:1-relation between "AMBER" chemical shifts and "CVFF" chemical shifts (Fig. 49). Especially for the methyl region a considerable scatter is observed. Thus, for finer details, the pecularities of the force field might be important. In fact, the correlation of the chemical shift with the lengths of the adjacent carbon bonds (cf. Fig. 45) is much weaker in the AMBER geometries. Again, to get rid of these uncertainties, reliable ab initio geometries are needed which will become available with the advent of more powerful computers.

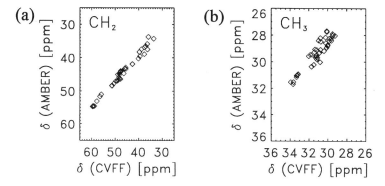

Fig. 49. The chemical shift obtained with AMBER geometries is well correlated with shifts relying on CVFF optimization for both the methylene (**a**) and the methyl (**b**) carbon. The considerable scatter implies that finer details may well depend on the force-field parametrization

5.3
Poly(butadiene) and Poly(isoprene)

1,4-poly(butadiene) (PBD) and 1,4-poly(isoprene) (PIP) are simple unsaturated polymers (Fig. 7) that permit one to investigate whether the presence of double bonds can be handled by our method. As the systems are very much alike, we will focus mainly on PBD and refer only occasionally to PIP. The statistical RIS-models for these systems [21, 22] assume six torsional states (*trans* (0°), *gauche* (±120°), *skew* (±60°), *cis* (180°)). The RIS models show some pecularities: They include explicit energy terms for triplets of dihedral angles, and the segments in between two double bonds are effectively decoupled from the rest of the molecule. The method, especially Eqs. (2.10), (2.12), has to be adapted to this special situation. To save computer time, it is advisable to choose a separate model molecule for the saturated and unsaturated carbons in PBD (Fig. 50.a/b), respectively. For the experimental measurements, a PBD sample with a *cis/trans*-ratio of 49:51 was used (^{13}C solution NMR); the calorimetric glass point was $T_g \approx 180$ K (DSC). The solid state NMR measurements were performed on a BRUKER ASX 500 at $T = 190$ K near T_g; no indications for motional averaging were found. For the simulation, 19 different geometries for model molecule A, and 27 conformations for model molecule B were considered with an overall probability of > 0.95 each. The basis II' has been used for quantum chemistry calculations. Figure 50.c/d shows the experimental spectrum in comparison with simulation. The *cis/trans*-splitting in the aliphatic region is well reproduced by the simulation whereas the unsaturated carbon is broader in the simulation. This finding is confirmed by the results of the solution spectrum. The *cis/trans*-splitting in CH$_2$ is 5.2 ppm experimentally at $T = 333$ K [112], and the simulation finds 5.7 ppm in the basis DZ and 6.3 in the basis II'. The splitting of the unsaturated carbon is only 0.6 ppm experimentally, whereas quantum chemistry finds 2.9 ppm (DZ) and 3.6 ppm (II') respectively. In contrast to the methylene group, the unsaturated carbons are only poorly described by the simulation. This finding might be an indication for shortcomings of the quantum chemistry IGLO calculations. In fact, comparable deviations from experimental *cis/trans*-splittings have been reported in the literature, e.g. for the small compounds 2-butene and 1,2-difluor-ethene [113]. Another unusual pattern is the better agreement in the small basis DZ compared with the larger basis II'. The problems on the quantum chemical side are well illustrated by Fig. 51. Here, the chemical shifts obtained with II' are plotted vs. DZ values. For both quantum chemistry calculations, the same set of geometries was used. Although the two sets of chemical shifts are still nicely correlated, the scatter is considerably larger than in the unsaturated systems PP, PE, and PIB. One possible explanation for the problems of

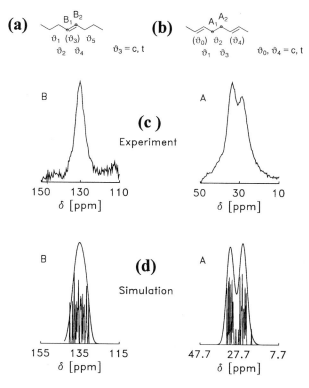

Fig. 50. For PBD, it is more practical to choose two different model molecules for the saturated and doubly bonded carbon, respectively. The number of relevant geometries and the overall size of the model molecule are considerably reduced by this procedure (**a, b**). Experimental (**c**) and simulated (**d**) spectrum for both the doubly bonded carbon B (*left*) and the saturated carbon (*right*) in PBD. The experimental spectrum has been recorded at $T = 190$ K, approximately the glass transition temperature. The *fine lines* symbolize the contribution of the various conformations, whereas the *bold line* denotes the overall simulation including the Gaussian broadening function. For the doubly bonded carbon, the simulation overestimates the width of the resonance. The *trans/cis* splitting of the saturated carbon A is well reproduced

the quantum chemistry calculations might be the larger influence of electron correlation in these unsaturated, "electron dense" systems. This would call for advanced quantum chemistry methods which include correlation effects either in a perturbation-theoretical treatment (GIAO-MBPT) [59, 60] or by a multi-configurational technique (MC-SCF IGLO) [40, 42]. Apart from this, the geometries could be partly defective due to a poor parametrization of the force field.

The *cis/trans*-splitting is essentially a γ-*gauche* effect enforced by the double bond (Fig. 52): The correlation of δ_{A_2} with $|\vartheta_1| + |\vartheta_4|$ is very much

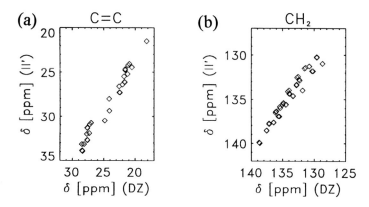

Fig. 51. The chemical shift obtained with the large basis II' is correlated with the DZ chemical shifts, though the correlation is appreciably worse than in aPP, PE, or PIB. The correlation for the saturated carbon (**b**) is better than the correlation for the doubly bonded carbon (**a**). This might be due to shortcomings of quantum chemistry in the unsaturated system PBD

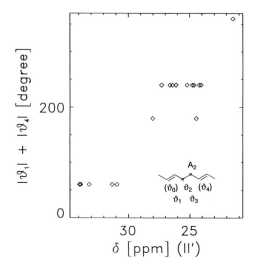

Fig. 52. The *trans/cis*-splitting of the unsaturated carbon in PBD is comparable to the usual *γ-gauche* effect: the chemical shift is correlated with $| \vartheta_1 | + | \vartheta_4 |$ as in aPP, PE, or PIB. There is no indication of a new mechanism to bring about the *cis/trans*-splitting. Different from fully saturated systems, the *cis/trans*-splitting is observable in solution as well, since the double bonds are either *cis* or *trans* and do not undergo *cis/trans*-transitions

alike the plot obtained for aPP, PE, and PIB. There is no indication of a separate mechanism providing the *cis/trans*-splitting. For PIP, the results are quite similar, but less easily interpreted due to the larger number of different

Table 9. *Cis/trans*-splitting $\Delta\delta = \delta_{cis} - \delta_{trans}$ for the different positions in 1,4-poly(isoprene). All values in ppm. The labels of the different sites are indicated in the sketches of the model molecules. For 1,4-poly(isoprene) it is convenient to employ different model molecules for the different sites

Position	Exp. [112] 298 K	IGLO (DZ) 300 K	IGLO (II') 300 K
A	not resolved	0.2	0.5
B	7.6	7.7	8.9
C	not resolved	0.6	1.5
D	0.9	1.7	2.5
E	7.6	4.4	4.1

sites and partially overlapping resonances. As in PBD, two model molecules were chosen as a basis for the simulation. We confine ourselves to present the results for the *cis/trans*-splitting in solution (Table 9).

An MO analysis reveals a larger electron delocalization in systems with multiple bonds. In contrast to saturated compounds, remote molecular orbitals (other than the MOs of the adjacent bonds) have appreciable contributions of > 1 ppm.

The results for the unsaturated systems PBD and PIP are less easily assessed than for aPP, PE, and PIB. The simulation results are partly in good agreement with experiment, and partly defective. Especially for the unsaturated carbons, large deviations from experimental data are found. To identify the origin of the difficulties (correlation effects ?, force field ?), and for a final decision on the validity of the simulation method for this class of systems, additional investigations will be necessary.

5.4
Poly(vinyl chloride)

Poly(vinyl chloride) (PVC) is a simple example of a polymer system that contains a hetero atom in addition to carbon and hydrogen (Fig. 7); it can be thus checked whether the method for the ab initio spectral simulation is reliable beyond hydrocarbons.

The PVC sample chosen as an experimental reference has a calorimetric glass transition temperature of $T_g = 356$ K and was purchased from Polyscience. The share of racemic diads was determined by solution ^{13}C -NMR

($p_r = 0.58$), and small deviations from ideal Bernoullian statistics were observed.

Remarkably, atactic PVC can exhibit substantial crystallinity [114, 115]. In our sample, crystallinity was checked by wide-angle X-ray scattering (WAXS) and found to be low ($< 15\%$).

For the conformational statistics we rely on a model developed by Williams, Flory, and Mark [116, 117, 118] which assumes three states (*trans t* ($0°$), *gauche g* ($+120°$) and $\bar{g}$ ($-120°$)). Note that in PVC, pseudoasymmetric carbons are present, and the special convention for the sense of rotation of the dihedral angles must be obeyed. A recent MD simulation [119] is in good agreement with the rotational isomeric state model (fraction of *trans* states, characteristic ratio).

The procedure for the spectral simulation follows closely the example of aPP. A tetramer is used as model molecule which allows the discrimination of tetrad sequences (Fig. 53.a). We are mainly interested in the methyl position A and confine ourselves to control the dihedral angles ϑ_1, ϑ_2, ϑ_3, ϑ_4 and the relative chiralities x_1, x_2, x_3; the exterior dihedral angles ϑ_0 and ϑ_5 are left undefined. The position of the γ-neighbors of the tertiary carbons B_1 and B_2 is thus not fully determined.

As in aPP, the symmetry element of changing the read direction of the segment, viz.

$$(\vartheta_1, \vartheta_2, \vartheta_3, \vartheta_4; x_1, x_2, x_3) \longrightarrow (\vartheta_4, \vartheta_3, \vartheta_2, \vartheta_1; x_3, x_2, x_1) \tag{5.6}$$

is used to reduce the number of relevant geometries. For the calculation, the conformational statistics at $T = 360$ K $= T_g$ was used; 36 geometries with an overall probability of > 0.85 were taken into account. The configurational statistics was assumed to be perfectly Bernoullian with the experimental *racemic* probability $p_r = 0.58$. Thus the observed deviations from the ideal Bernoullian behaviour were ignored. Geometry optimization utilized CVFF force-field minimization, and the IGLO computations were performed with the atomic basis DZ. As usual, the simulated spectrum is obtained by Eq. (3.16) (Fig. 53). The experimental spectrum displays two rather featureless Gaussian-shaped resonances for the two different sites. The simulation is in good agreement with experiment, the widths of the peaks are well reproduced. The small high field shoulder of the methylene resonance is present only in the simulation, and absent in the experiment. A more detailed analysis of the simulation reveals that this shoulder is connected with the assumed stereo statistics; most likely the discrepancy between experiment and theory is due to the slightly oversimplified Bernoullian statistics used in the model.

The methylene carbons display a quite pronounced γ-*gauche* effect (Fig. 54.a) which is predominantly steric in origin. The latter assertion can

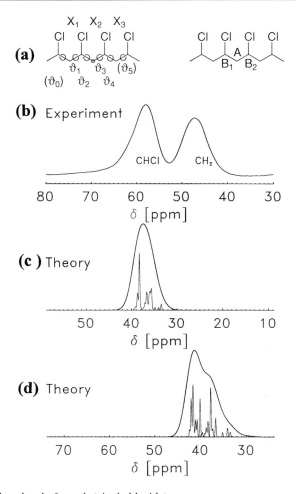

Fig. 53. Model molecule for poly(vinyl chloride).
The configurational degrees of freedom (x_i), the conformational parameters (ϑ_i) and the labels of the central atoms are depicted (**a**).
Experimental CP/MAS spectrum of PVC (**b**) (recorded at room temperature well below the calorimetric glass transition temperature) in comparison with simulation results for CHCl (**c**) and CH$_2$ (**d**). The *fine lines* denote the contributions of the various geometries, according to Eq. (3.15), whereas the *bold lines* symbolize the overall simulation including the Gaussian broadening function, Eq. (3.16). The width of both resonances is well met. The small shoulder in the simulated methylene resonance is most likely due to a defective configurational statistics

be inferred from the good correlation of the chemical shift with the distance of the γ-neighbors (Fig. 54.b).

The tertiary carbons have four carbon γ-partners and a simple γ-*gauche* effect is not to be expected, and not observed in fact.

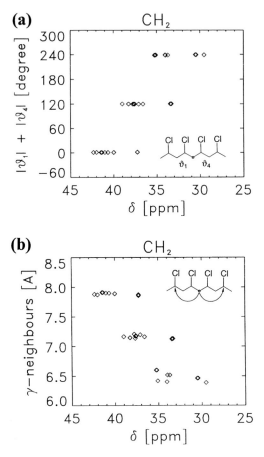

Fig. 54. The methylene carbon in PVC displays a γ-*gauche* effect, i.e. a correlation with $|\vartheta_1| + |\vartheta_4|$ (**a**). The observed scatter within the horizontally aligned clusters betrays the importance of geometric parameters other than ϑ_1 and ϑ_4. The γ-*gauche* effect partly stems from a steric interaction, as can be inferred from the correlation of δ with the γ-neighbor distance (**b**)

PVC contains pseudoasymmetric carbons and a configurational splitting in solution is observed. Unfortunately, the signals in the methylene region are dominated by solvent effects. Neither the traditional γ-*gauche* method nor the ab initio technique are able to make realistic predictions for the methylene region. For the tertiary carbons, the situation is different, and both methods are in good agreement with experiment (Table 10).

One of the most important results of this section refers to the influence of the atomic basis. It is not clear *a priori* whether a small basis like DZ is still sufficient to treat a system containing a comparably heavy hetero atom like chlorine. Fortunately, this is no obstacle in practice (Fig. 55). As in the

Table 10. Configurational splitting of the tertiary carbon in poly(vinyl chloride). The value for the *mm*-diad has been set to 0 ppm. The experimental values have been determined graphically from [133] with an accuracy of ≈ 0.1–0.2 ppm

Diad	Exp. [133]	Exp. [112]	IGLO (DZ)	emp. method [133]
mm	0.0	0.0	0.0	0.0
mr	0.85	1.0	0.73	0.85
rr	1.65	1.9	1.97	1.55

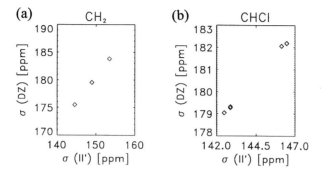

Fig. 55. The presence of the chlorine atom does not affect the IGLO results obtained with the DZ atomic basis. For both the methylene carbon (**a**) and the tertiary carbon (**b**), the DZ shift is well correlated with the shifts obtained with the larger basis II'. As in the other systems, the time-consuming basis II' may be replaced by DZ for relative effects like the conformational broadening and the configurational splitting

other systems, the chemical shift calculated with the small atomic basis DZ is well correlated with the chemical shifts obtained with the larger basis II'. Only the absolute offset is not correctly predicted by the small basis. Thus, geometry-induced variations can be reliably obtained with a small atomic basis even for a system like PVC. Apparently, hetero atoms in themselves are no obstacle to the ab initio spectral simulation of polymer systems.

5.5
Poly(methyl methacrylate)

5.5.1
Calculations and Simulated Spectra

Poly(methyl methacrylate) (PMMA) is a vinyl polymer with two sidegroups, one of which is extended (COOCH$_3$) and has internal degrees of freedom (Fig. 7). The geometry of a PMMA segment is determined not only by the main chain dihedral angles ϑ_i and the chirality x_i of the quaternary carbons, but also by the orientation χ_i of the extended sidegroups with respect to the

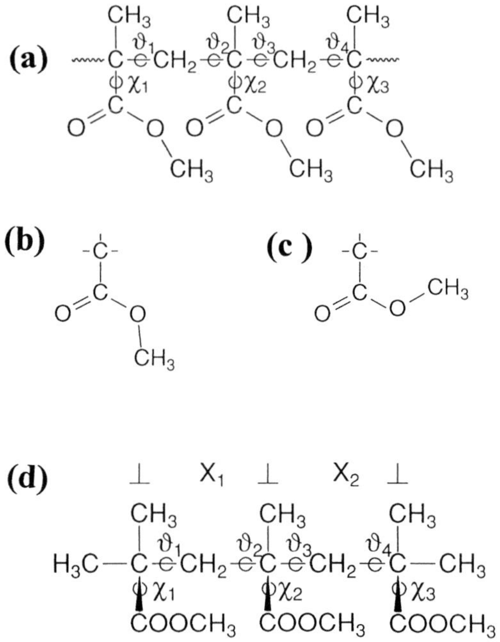

Fig. 56. Segment of an extended PMMA chain (**a**). In PMMA, not only the backbone dihedral angles ϑ_i are relevant, but also the orientation χ_i of the sidegroups with respect to the main chain. The COOCH$_3$ sidegroup is partially conjugated; therefore only two different geometries (**b, c**) are possible. The geometry (**c**) can be ruled out by general arguments and comparison with other systems. The model molecule for PMMA is a suitably saturated trimer (**d**). The relevant degrees of freedom include the configuration (x_1, x_2), the main chain conformation $(\vartheta_1, \vartheta_2, \vartheta_3, \vartheta_4)$ and the orientation of the sidechains (χ_1, χ_2, χ_3). This model molecule is not suitable for investigating the methylene group CH$_2$ (see text)

main chain (Fig. 56.a). The internal degrees of freedom of the sidechain itself need not be considered separately, as due to partial conjugation the methyl ester group is planar with only two possible geometries one of which may be ruled out by theoretical and experimental findings [120] (Fig. 56.b/c). The overall geometry of the trimer (Fig. 56.d) which we select as a model molecule for the quantum chemical calculations, is thus given by the sequence

$$(\vartheta_1, \vartheta_2, \vartheta_3, \vartheta_4; \chi_1, \chi_2, \chi_3; x_1, x_2) \ . \tag{5.7}$$

We will be interested only in the central carbon atoms C$_\alpha$ (quaternary), CH$_3$ (methyl sidegroup), and COO (carboxylic group). The methylene and the methyl ester (OCH$_3$) peaks overlap experimentally. Therefore, they are both omitted from the present analysis. For our purposes, we consider the two terminal CH$_3$ groups (which symbolize the continuation of the chain) as

Table 11. Rotational Isomeric States assumed in the model [24] for poly(methyl methacry-late). In this pseudoasymmetric polymer, the sense of rotation depends on the chirality of the asymmetric centers

Name	t_-	t_+	g_-	g_+	$\bar{g}_-$	$\bar{g}_+$
ϑ	$-20°$	$+10°$	$+100°$	$+125°$	$-125°$	$-100°$

distinguished from the methyl sidegroups; the two terminal C_α-atoms are still chiral then.

This system is far more complicated than PP or PE. Special difficulties stem from the numerous degrees of freedom and the large steric interactions. As a consequence, the appropriate RIS model for this system is still debated. En-ergy maps of two consecutive dihedral angles may be misleading as they tend to underestimate the impact of the other parameters that have *to be held fixed in a two-dimensional representation. Among the various approaches [120, 121, 24], we adopt the six-state model of Vacatello and Flory [24], as it is – in spite of some yet unexplained features – supported by neutron scat-tering data [122, 123, 124] and recent molecular dynamics simulations [125]. It should be emphasized, however, that an RIS model for the complicated system PMMA is bound to be less accurate than a corresponding model for a simpler polymer.

The basic assumptions of the model [24] are:

1. The six states of the main chain dihedral angles take the values given in Table 11. As in aPP and PVC, the sense of rotation depends on the chirality of the substituted carbon atom.
2. For the orientation of the sidechains χ_i, only the two values $\chi_i = 0°$, $180°$ are permitted.
3. The statistical weights depend only on the main chain dihedral angles ϑ_i and not on the sidegroup orientation χ_i. This is a simplification, chosen in [24] to obtain a manageable model.
4. The bond angles do depend on the dihedral angles in a characteristic way. In our approach, this feature is automatically introduced by the empirical force field.

As experimental references, a preferentially isotactic sample ($p(mm) > 0.95$) and a highly syndiotactic sample ($p(rr) = 0.62$, $p(mr) = 0.30$, $p(mm) = 0.08$) were selected. Crystalline phases were excluded for both samples by WAXS. Calorimetric glass transition temperatures were $T_g = 330$ K for the isotactic sample, and $T_g = 400$ K for the syndiotactic modification, respectively.

For the ab initio spectral simulations, the many degrees of freedom to-gether with the high glass transition temperature are a special nuisance, as the number of relevant geometries is largely increased. For the model molecule, $2^2 \cdot 6^4 \cdot 2^3 \approx 41{,}000$ different geometries are possible. To achieve an overall probability of > 0.85 at $T = 400$ K, still more than 1000 conformations must be considered. Facing a computation time of ≈ 27 h (SGI IRIS INDIGO) for a single geometry, a full simulation with all these geometries is unattainable at present. As for the other systems, the number of geometries can be considerably reduced by making use of the "chain-reverse" symmetry

$$(\vartheta_1, \vartheta_2, \vartheta_3, \vartheta_4; \chi_1, \chi_2, \chi_3; x_1, x_2) \; . \longrightarrow (\vartheta_4, \vartheta_3, \vartheta_2, \vartheta_1; \chi_3, \chi_2, \chi_1; x_2, x_1) \; . \quad (5.8)$$

But still the number of relevant geometries is very large and must be reduced by an additional simplification. Considering the approximate character of the underlying statistical model and being interested only in the central carbon atoms, we may neglect the impact of the orientation of the remote side-groups (χ_1, χ_3) on the chemical shift in the central part of the molecule. The error introduced by this simplification was checked by a separate calculation (6 geometries) and shown to be less than an average 1.4 ppm for each group. For CH_3, C_α, and OCH_3, the maximum error was below 1 ppm. Combining symmetry and the neglect of χ_1 and χ_3, the number of relevant geometries is reduced to about 100 which can be treated on a workstation cluster. The geometries were optimized with the empirical CVFF force field, and the subsequent IGLO calculations were performed with the DZ atomic basis. The simulated spectra were obtained by applying Eq. (3.16) with the experimental configurational statistics. The width of the convolution function was 1.5 ppm as for the other polymers. The conformational statistics at the respective glass transition temperatures (400 K for sPMMA, 330 K for iPMMA) were used.

In contrast to the examples discussed so far, the overall agreement with the experimental spectra is generally poor, as can be seen from the simulations of the methyl sidegroup (Fig. 57) and the carboxylic group (Fig. 58) which are presented here as examples. The shape of the spectra is not met and the simulation seems to overestimate the width of the resonances. In this system with no well-defined crystalline modifications being easily available, it is a particular drawback that the method is unable to predict the absolute chemical shifts correctly. Note however, that the error in the absolute shift (≈ 28 ppm for COO, ≈ 4 ppm for CH_3) may be quite accurately estimated by comparison with the solution results (Sect. 5.5.2, Fig. 63).

Thus, at first sight, the agreement between experiment and simulation is disappointing. In our understanding, however, this is rather remarkable as PMMA is a special case. X-ray studies [126, 127] and recent NMR experiments [128, 129, 130, 131] have found evidence for residual local conformational order in the solid glassy state in contrast with the usual random coil picture of amorphous polymers. An RIS model such as [24] does not take into account special ordering effects brought about by neighboring chains. For an RIS-

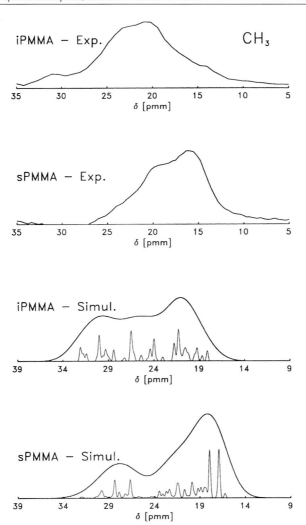

Fig. 57. From *top* to *bottom:* experimental spectrum of almost ideally isotactic PMMA ($p(mm) > 95\%$), experimental spectrum of preferentially syndiotactic PMMA ($p(rr) = 62\%$, $p(mr) = 30\%$, $p(mm) = 8\%$), simulation for ideally isotactic PMMA, simulation of PMMA with the configurational statistics of the experimental preferentially syndiotactic sample. Only the methyl region of the spectra is shown. Compared with aPP (Fig. 16) the agreement between experiment and theory is poor. Especially for the syndiotactic sample the simulated line shape is broader than the experimental one, which is indicative of some residual order in the amorphous glassy state. The *fine lines* show the contributions of the various conformations and the *bold lines* the overall simulation with the Gaussian broadening function s_b

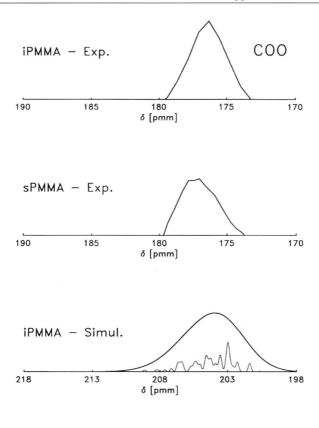

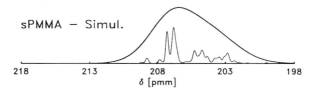

Fig. 58. Experimental and simulated spectrum for the carboxylic carbon in PMMA. For more details see the legend of Fig. 57. As for the methyl group, the simulated spectrum is broader than the experiment which is indicative of some residual order in the polymer glass

model neither the solvent (in a dilute solution) nor the other chains (in the melt or the glass) have an impact on the *average* geometry of a single chain. Their influence is thought to be shielded on average. The physical situation in the PMMA glass is thus different from the model assumptions. The poor agreement of experimental solid state NMR spectra and simulation thus further points to deviations of the conformational statistics from that in a random coil. This assertion refers *only* to a local length scale of a few repeat units, and the analysis of the neutron scattering data, which probe the molecular organization on a larger length scale, is not contradicted. A closer look on the situation reveals that a locally ordered PMMA chain (as evidenced by NMR) is, in fact, not in conflict with a solution-like global random coil (as can be seen from the characteristic ratio that is similar to the ϑ-solution value [16, 17]). The reason for this is twofold. On the one hand, experimentally, the characteristic ratio is sensitive mainly to the loss of correlation on the extended length scale of the radius of gyration R_g irrespective of local effects. On the other hand, the characteristic ratio inferred from the RIS model [24] is influenced not only by conformation (i.e. dihedral angles) but also by bond angles [24, p. 413]. Thus there is no *strict* relation between local conformation and the experimental characteristic ratio, and a local conformational ordering may well be incorporated into the general framework of a globally disordered amorphous polymer.

In view of the experimental findings [126, 127, 128, 129, 130, 131] it is clear now why the simulation yields broader peaks than experiment. In reality, the disorder is reduced due to partial conformational "alignment", and consequently, the peaks are narrower in experiment.

Shortly, it will become clear, that the results for the configurational splitting in solution are in very good agreement with experiment. From this, we conclude that neither the single chain statistical model in itself nor the quantum chemistry calculations suffer from severe drawbacks, and that the discrepancy seen in Figs. 57, 58 indicates local conformational order in the glassy state and is not an artifact introduced by the calculation. In view of the numerous approximations and limitations of the PMMA calculation, more definite statements are not justified and must be deferred to the future, when faster computers will become available.

5.5.2
Geometry Effects, Role of Atomic Basis, and Configurational Splitting

PMMA is a crowded system with large steric interactions; in this respect it is comparable to PIB. And like in PIB, the inhomogeneous broadening of the methyl sidegroup cannot be explained by a simple *γ-gauche* effect. There is no indication of a correlation of δ_{CH_3} with $| \vartheta_2 | + | \vartheta_3 |$, the added moduli of the *γ-gauche* dihedral angles (Fig. 59.a) although the chemical shift of the methyl group is related to the distance of its main chain γ-neighbors (Fig. 59.b). As in PIB, the reason for this unusual behaviour stems from the widening of the bond angles, if one or both of the adjacent dihedrals is in a *gauche* state (Fig. 60). The actual *shape* of the molecule cannot be described by the dihedrals alone. This is another breakdown of the basic assumptions

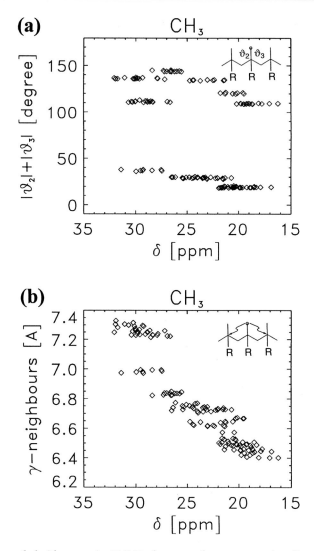

Fig. 59. The methyl sidegroup in PMMA does not show a γ-*gauche* effect; there is no correlation of δ with $|\vartheta_2| + |\vartheta_3|$ (**a**), although the chemical shift is correlated with the γ-neighbor distance (**b**). This apparent contradiction reveals that in the crowded system PMMA, the dihedral angles are no longer the proper parameters to describe the overall geometry

of the phenomenological γ-*gauche* concept, which will have implications for the prediction for the configurational splitting in solution (*vide infra*).

The quaternary carbon C_α has seven (!) γ-partners, and the steric situation does not change much if ϑ_1 or ϑ_4 are set to another rotational isomeric state.

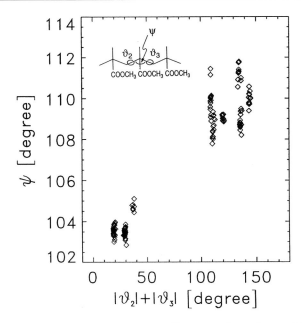

Fig. 60. The large steric interactions in PMMA effectuate a dependence of the bond angle ψ on the adjacent dihedral angles ϑ_2 and ϑ_3. The bond angle is widened if either ϑ_2 or ϑ_3 is in a *gauche* position. By this mechanism, the direct influence of the dihedral angles on the γ-neighbor distance is overcompensated, and the phenomenological understanding of the γ-*gauche* effect experiences a breakdown

As a kind of replacement, there is a remarkable correlation of δ_{C_α} with the distances of the atoms of the extended sidegroup (Fig. 61).

For the carboxylic carbon, it is even harder to find a sensible geometry correlation. Neither $\mid \vartheta_2 \mid + \mid \vartheta_3 \mid$ nor the distances of the γ-neighbors seem to be of particular importance for the chemical shift. Apart from a very weak correlation with the distance of the neighboring carboxylic carbons, no special geometry parameter has a systematic impact on the chemical shift.

The spread of the methyl ester carbon OCH_3 is too small for a useful analysis, and the methylene group CH_2 is not included in the present analysis, as it is not properly represented in the small model molecule.

To conclude, in PMMA, we experience an almost complete breakdown of the traditional ideas on how the geometry should influence the chemical shift. PMMA is too complicated for a simple phenomenological parametrization.

To assess the influence of the small atomic basis on the results of the simulation, five geometries were recalculated with the larger basis II'. For CH_3 and C_α, the chemical shifts obtained with the larger basis are well correlated with the DZ results (Fig. 62.a,b). For these sites at least, the small basis DZ is sufficient to calculate the geometry-induced variations. The absolute

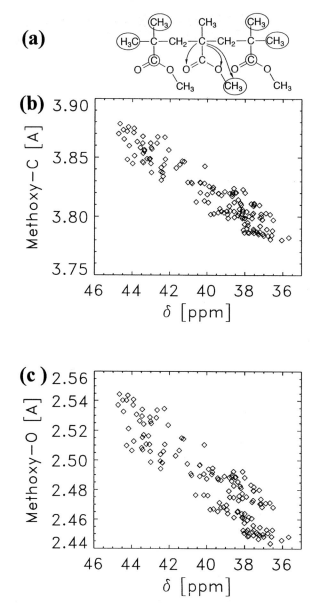

Fig. 61. The central quarternary carbon in PMMA has seven γ-neighbors (**a**, *in circles*). Because of this, a simple geometry dependence cannot be expected. In fact, the only appreciable geometry correlation is with the distance of the methoxy carbon (**b**) and the methoxy oxygen (**c**) of the COOCH₃ sidegroup

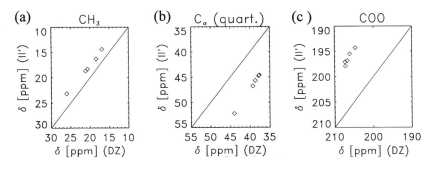

Fig. 62. In PMMA, the chemical shift obtained with the large basis II' is well correlated with the ordinary DZ chemical shifts for the methyl carbon (**a**), the quarternary carbon (**b**), and the carboxylic carbon (**c**). For the latter, a considerable scatter is observed. Still, the basis DZ is useful in computing relative chemical shift, although the absolute values are not properly reproduced

chemical shift is not well reproduced, as usual. For the carboxylic carbon, the observed scatter is larger, but still small enough for the needed accuracy. Correlation effects might be important for the carboxylic carbon. It should be noted, that for all sites the variation of the chemical shift is even larger in the more realistic basis II'. Therefore, the conclusions put forward in Sect. 5.5.1 (interpreting the solid state NMR as indicating local conformational order in the glassy state) are strengthened by the calculations with the larger basis set.

PMMA has pseudoasymmetric centers, and its solution spectrum displays a very large configurational splitting. Considering the complete absence of a γ-*gauche* effect in terms of dihedral angles, any attempt to explain the configurational splitting by the traditional method is bound to fail. This was indeed reported by Tonelli [132], who already suggested that the widening of the bond angles might be responsible for his findings – a conjecture now substantiated by the ab initio calculations. The ab initio simulation of the configurational splitting, Eq. (3.17), does not rely on special assumptions on the geometry dependence of the chemical shift. It just sums up the contributions of the various conformations, irrespective of the way the chemical shift responds to the change in geometry. The results are in excellent agreement with experimental data (Fig. 63, experimental data taken from [112]). Both the size of the splitting and the assignments are correct. In contrast to this, the maximum splitting found by the empirical method (C_α: 0 ppm, CH$_3$: 1.2 ppm, COO: 0.9 ppm) were far too small, and for the carboxylic carbon even the order of the subresonances was reversed.

The good agreement of simulation and experiment for the configurational splitting is not in conflict with the discrepancies in the solid state spectrum.

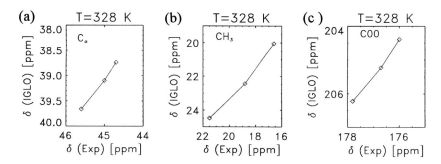

Fig. 63. The configurational splitting for C_α, CH_3, and COO carbons as predicted by IGLO are in excellent agreement with experimental data, taken from [112]. For PMMA, the ab initio method could thus resolve a puzzle which was left unexplained by the empirical method

There, special packing mechanisms lead to a partial chain order, that is not accounted for in a single chain RIS model. In solution, the random coil model is valid on average even on a local length scale, and hence good agreement with experiment is found.

6
Conclusion

In this study we have presented the first simulations of solid state NMR spectra of polymers that are based on ab initio quantum chemistry. By combining statistical conformational models, empirical force-field optimization of geometries, and the "Individual Gauges for Localized Molecular Orbitals" (IGLO) method for the calculation of the chemical shift, it is possible to reproduce the shape and the width of the experimental ^{13}C solid state NMR spectra.

By relating calculated chemical shifts and the underlying geometry it is possible to identify the geometries contributing to the subresonances of the experimental spectra. Solid state NMR together with advanced theoretical methods may be thus used as a source of structural information for glassy amorphous polymers. This is especially valuable, as NMR focuses on a strictly local length scale of a few bonds and may thus complement the scattering methods which probe amorphous polymers on a larger scale and yield global parameters that are in turn unattainable to NMR.

For simple systems, the *γ-gauche* effect is essentially confirmed, i.e. the chemical shift is related to the value of the second next dihedral angle (counted from the atom in question) in a systematic way. It becomes clear, however, that other geometrical parameters play an important role as well. The conception of a *γ-gauche* effect depending exclusively on the respective dihedral angles is oversimplified and has to be modified for a more detailed treatment. In compounds with large steric hindrance (poly(isobutylene), poly(methyl methacrylate)), the widening of the bond angles may overcompensate the effect of the dihedral angles, and a proper *γ-gauche* effect (understood as the influence of the dihedral angles) is no longer discernible. The geometry dependence of the chemical shift is more complex in these cases. Sometimes, unexpected correlations with bond lengths or bond angles are found (poly(isobutylene), poly(ethylene)).

An MO analysis reveals the strictly local character of the chemical shift which stems almost exclusively from the electrons of the inner shell of the atom itself and the adjacent bonds.

Moreover, the configurational splitting in solution may be simulated as well. Quantum chemical calculations can thus provide a justification for the empirical *γ-gauche* method which uses a simple increment system to predict the chemical shift. In some systems (e.g. poly(methyl methacrylate)), where a breakdown of the empirical method is experienced, quantum chemistry may successfully predict the splitting as it does not rely on special assumptions. For routine analysis, the higher resolution and the much lower computation time favor the traditional phenomenological approach. In these cases, quantum-chemistry based simulations should be regarded as a complement rather than a replacement.

Using the capacity of IGLO to calculate the full chemical shift tensor, first tentative results on the chemical shift anisotropy in atactic poly(propylene) indicate that the tensorial character of the chemical shift may be used as source of structural information, too.

To check the range of applicability, the method has been applied to a number of polymer systems belonging to different structural classes. In saturated polyolefines (atactic poly(propylene), poly(ethylene), poly(isobutylene)) no problems are encountered, and the experimental findings are usually well reproduced by the simulation. For unsaturated systems (poly(butadiene), poly(isoprene)), some unusual findings suggest the need for improvements on the quantum chemistry side. Hetero atoms are not a problem in themselves, and the method may be applied without modifications (poly(vinyl chloride), poly(methyl methacrylate)). Complex systems with many degrees of freedom, a high glass transition temperature and many relevant conformations (such as poly(methyl methacrylate)) require an especially high computing power and will be fully accessible only in the future.

The method may be used to discriminate between different statistical models by comparing the respective simulation result with experiment (poly(isobutylene)), to check the validity of otherwise established models on a local length scale (atactic poly(propylene)) and to draw conclusions from apparent deviations between experiment and simulation (poly(methyl methacrylate)).

Additional calculations show that the size of the model molecule for the quantum chemistry calculations may be effectively restricted to the γ- and δ-neighbors of the atoms in question, and that only minor errors are introduced by end-group effects. The necessary computing time is greatly reduced by the fact that even moderate atomic basis sets (of *double zeta* quality) may be used for the calculation of relative, geometry-induced effects. As the chemical shift is very sensitive even to small variations of the geometry, the choice of an appropriate force field is crucial to the reliability of the results. Fortunately, basic findings (like the shape of the simulated spectrum and the *γ-gauche* effect) are rather insensitive to parametrization, as demonstrated by a parallel simulation of poly(isobutylene) with CVFF and AMBER geometries. Finer details, however, *do* depend on force-field pecularities, and further research is needed to clarify this point.

The method presented in this study consists of the combination of three quite independent building blocks:

- a statistical model to obtain the probabilities of the relevant geometries,
- a method to generate the actual coordinates of the model molecule (a force field in this study),
- an ab initio scheme to calculate the chemical shift for the given geometries.

When selecting the building blocks, tribute has been paid to the availability of statistical models and the power of present-day computers. With the advent of more powerful computers and novel quantum chemistry techniques, it will be possible to replace the force-field optimization by advanced ab initio approaches and the Hartree-Fock IGLO method used here by schemes that include electron correlation.

Most important from the polymer science point of view, it is already possible today to replace one RIS model by another or even a more sophisticated approach. To be used as the basis for a spectral simulation, a statistical model must meet essentially two requirements. It must be sufficiently precise on a strictly molecular level, and the number of relevant geometries should not exceed a few hundred. The latter requirement naturally depends on the available computer capacity.

To conclude, it is believed that ab initio quantum chemistry methods in the spirit of the approach presented in this paper will play an increasing role in the investigation of structural and geometrical properties of polymer glasses. They provide tools for quantitative analysis of NMR spectra to yield structural information on a strictly local length scale. Solid state NMR can thus complement the scattering methods which probe the samples mainly at greater distances.

References

1. Campbell D., White J. R. (1989) Polymer Characterization. Chapman and Hall, New York

2. Spells S. J. (ed). (1994) Characterization of Solid Polymers. Chapman and Hall, New York

3. Schmidt-Rohr K., Spiess H. W. (1994) Multidimensional Solid State NMR and Polymers. Academic Press, New York

4. Bovey F. A. (1982) Chain Structure and Conformation of Macromolecules. Academic Press, New York

5. Flory P. J. (1969) Statistical Mechanics of Chain Molecules. Interscience Publishers, New York

6. Mattice W. L., Suter U. W. (1994) Conformational Theory of Large Molecules. J. Wiley and Sons, New York

7. Conradini P., Guerra G. (1992) Adv. Polym. Sci. 100:183

8. Tonelli A. E. (1989) NMR Spectroscopy and Polymer Microstructure: The Conformational Connection. VCH, Weinheim, New York

9. Born R., Spiess H. W., Kutzelnigg W., Fleischer U., Schindler M. (1994) Macromolecules 27:1500–1504

10. Born R., Spiess H. W. (1995) Macromolecules 28:7785–7795

11. Auriemma F., Born R., Spiess H. W., De Rosa C., Corradini P. (1995) Macromolecules 28:6902–6910

12. Flory P. J., Sundararajan P. R., DeBolt L. C. (1974) J. Am. Chem. Soc. 96:5015–5024

13. Richert R., Blumen A. (eds). (1994) Disorder Effects on Relaxational Processes. Glasses, Polymers, Proteins. Springer, Berlin, Heidelberg, New York

14. Elias H. G. (1977) Macromolecules I. John Wiley, New York

15. Fischer E. W., Wendorff J. H., Dettenmaier M., Lieser G., Voigt-Martin I. (1974) Polymer Preprints 15(2):8–13

16. Kirste R. G., Kruse W. A. (1973) Z. Makromol. Chem. 162:299

17. Boothroyd A. T., Rennie A. R., Wignall G. D. (1993) J. Chem. Phys. 99:9136–9144

18. Zirkel A., Urban V., Richter D., Fetters L. J., S. Huang J., Kampmann R., Hadjichristides N. (1992) Macromolecules 25:6148–6155

19. Mehring M. (1983) Principles of High Resolution NMR in Solids. Springer, Berlin, Heidelberg, New York

20. Toda M., Kubo R., Saito N. (1992) Statistical Physics. Springer, Berlin, Heidelberg, New York. Second edition

21. Abe Y., Flory P. J. (1971) Macromolecules 4:219–229

22. Abe Y., Flory P. J. (1971) Macromolecules 4:230–238

23. Suter U. W., Saiz E., Flory J. P. (1983) Macromolecules 16:1317–1328

24. Vacatello M., Flory P. J. (1986) Macromolecules 19:405–415

25. Smith G. D., Yoon D. Y., Jaffe R. L. (1993) Macromolecules 26:5213–5218

26. Vacatello M., Yoon D. Y. (1992) Macromolecules 25:2502–2508

27. Raucchi R., Vacatello M. (1993) Makromol. Chem., Theory Simul. 2:875–888

28. Binder K. (ed). (1995) Monte Carlo and Molecular Dynamics Simulations in Polymer Science. Oxford University Press, New York

29. Bowen J. P., Allinger N. L. (1991) Molecular mechanics: The art and science of parametrization. In : Lipkowitz K. B., Boyd D. B. (eds) Reviews in Computational Chemistry. Vol 2. VCH, Weinheim, New York

30. Dinur U., Hagler A. T. (1991) New approaches to empirical force fields. In : Lipkowitz K. B., Boyd D. B. (eds) Reviews in Computational Chemistry. Vol 2. VCH, Weinheim, New York

31. Burkert U., Allinger N. L. (1982) Molecular Mechanics. Volume 177 of ACS Monographs. American Chemical Society, Washington

32. Price S. L. (1990) Towards realistic model intermolecular potentials. In : Catlow C. R. A., Parker S. C., Allen M. P. (eds) Computer Modelling of Fluids, Polymers, and Solids. NATO ASI Series. Kluwer Academic Publishers, Dordrecht, Boston, London, p 29–54

33. Biosym, Inc. DISCOVER. (1993)

34. Weiner S. J., Kollman P. A., Case D. A., Chandra Singh U., Ghio C., Alagona G., Profeta S., Weiner P. (1985) J. Am. Chem. Soc. 106:765–785

35. Weiner S. J., Kollman P. A., Nguyen D. T., Case D. A. (1986) J. Comp. Chem. 7:230–252

36. Kutzelnigg W. (1980) Isr. J. Chem 19:193–200

37. Schindler M., Kutzelnigg W. (1982) J. Chem. Phys. 76:1919–1933

38. Kutzelnigg W., Fleischer U., Schindler M. (1991) NMR Basic Principles and Progress 23:165–262

39. Kutzelnigg W. (1989) J. Mol. Struct.(Theochem) 202:11–61

40. Kutzelnigg W., Wüllen Ch. van, Fleischer U., Franke R., Mourik T. von (1993) The IGLO method. Recent developments. In : Tossel J. A. (ed) Nuclear Magnetic Shielding and Molecular Structure. NATO ASI Series. Kluwer Academic Publishers, Dordrecht, Boston, London, p 141–162

41. Meier U., Wüllen Ch. van, Schindler M. (1992) J. Comp. Chem. 13:551–559

42. Wüllen Ch. van, Kutzelnigg W. (1993) Chem. Phys. Lett. 205:563–571

43. Kutzelnigg W. (1975) Einführung in die theoretische Chemie. Volume 1 VCH, Weinheim, New York

44. Kutzelnigg W. (1978) Einführung in die theoretische Chemie. Volume 2 VCH, Weinheim, New York

45. McWeeny R. (1992) Methods of Molecular Quantum Mechanics. Academic Press, London. Second edition

46. Boys S. F. (1960) Rev. Mod. Phys. 32:296–299

47. Boys S. F., Foster J. M. (1960) Rev. Mod. Phys. 32:300–302

48. Boys S. F., Foster J. M. (1960) Rev. Mod. Phys. 32:303–304

49. Boys S. F., Foster J. M. (1960) Rev. Mod. Phys. 32:305–307

50. London F. (1927) Z. Naturwiss. 15:187

51. Pople J. A. (1962) J. Chem. Phys. 37:53–59

52. Pople J. A. (1962) J. Chem. Phys. 37:60–66

53. Hameka H. (1958) Mol. Phys. 1:203–215

54. Ditchfield R. (1972) J. Chem. Phys. 56:5688–5691

55. Wolinski K., Hinton J. F., Pulay P. J. (1990) J. Am. Chem. Soc. 112:8251–8260

56. Pulay P., Hinton J. F., Wolinski K. (1993) Efficient implementation of the GIAO method for magnetic properties: Theory and application. In : Tossel J. A. (ed) Nuclear Magnetic Shielding and Molecular Structure. NATO ASI Series. Kluwer Academic Publishers, Dordrecht, Boston, London, p 243–262

57. Hansen A. E., Bouman T. D. (1985) J. Chem. Phys. 82:5035–5047

58. Hansen A. E., Bouman T. D. (1993) Ab-initio calculation and analysis of nuclear magnetic shielding tensors: the LORG and SOLO approaches. In : Tossel J. A. (ed) Nuclear Magnetic Shielding and Molecular Structure. NATO ASI Series. Kluwer Academic Publishers, Dordrecht, Boston, London, p 95–116

59. Gauss J. (1993) J. Chem. Phys. 99:3629–3642

60. Gauss J. (1994) Chem. Phys. Lett. 229:198–203

61. Huzinaga S. (1965) J. Chem. Phys. 42:1293–1302

62. Randall J. C. (1977) Polymer sequence determination. Academic Press, New York

63. Zemke K., Schmidt-Rohr K., Spiess H. W. (1994) Acta Polymerica 45:148–159

64. Spera S., Bax A. (1991) J. Am. Chem. Soc. 113:5490–5492

65. Ando I., Saito H., Tabeta R., Shoji A., Ozaki T. (1984) Macromolecules 17:457–461

66. Jiao D., Barfield M., Hruby V. J. (1993) J. Am. Chem. Soc. 115:10883–10887

67. Jiao D., Barfield M., Hruby V. J. (1993) Magn. Res. Chem. 31:75–79

68. Kuruso H., Ando I., Webb G. A. (1993) Magn. Res. Chem. 31:399–402

69. Imashiro F., Masude Y., Honda M., Obara S. (1993) J. Chem. Soc. Perkin Trans. II 1535–1541

70. Sternberg U., Priess W. (1993) J. Mag. Res. A 102:160–165

71. Barfield M. (1993) J. Am. Chem. Soc. 115:6919–6928

72. Dios A. C. de, Pearson J. G., Oldfield E. (1993) Science 260:1491–1496

73. Dios A. C. de, Oldfield E. (1994) J. Am. Chem. Soc. 116:5307–5314

74. Tossel J. A. (ed). Nuclear Magnetic Shieldings and Molecular Structure (1993) NATO ASI Series. Kluwer Academic Publishers, Dordrecht, Boston, London

75. Imashiro F., Obara S. (1995) Macromolecules 28:2840–2844

76. Wagner G. (1990) Progress in NMR Spectroscopy 22:101–139

77. Sternberg U. (1993) The influence of structure and geometry on the ^{31}P, ^{29}Si, ^{13}C and ^{1}H chemical shifts. In : Tossel J. A. (ed) Nuclear Magnetic Shielding and Molecular Structure. NATO ASI Series. Kluwer Academic Publishers, Dordrecht, Boston, London, p 435–448

78. Grimmer A.-R. (1993) Shielding tensor data and structure: The bond-related chemical shift concept. In : Tossel J. A. (ed) Nuclear Magnetic Shielding and Molecular Structure. NATO ASI Series. Kluwer Academic Publishers, Dordrecht, Boston, London, p 191–202

79. Geen H., Titman J. T., Gottwald J., Spiess H. W. (1994) Chem. Phys. Lett. 227:79–86

80. Sommer W., Gottwald J., Demco D. E., Spiess H. W. (1995) J. Magn. Res. A 113:131–134

81. Geen H., Titman J. J., Gottwald J., Spiess H. W. (1995) J. Magn. Res. A 114:264–267

82. Gottwald J., Demco D. E., Graf R., Spiess H. W. (1995) Chem. Phys. Lett. 243

83. Gullion T., Schaefer J. (1989) Adv. Magn. Res 13:58–84

84. Bennet A. E., Griffin R. G., Vega S. (1994) NMR Basic Principles and Progress 33:1–77

85. Zemke K. Untersuchungen zur Konformation und Dynamik von amorphen Polymeren mit mehrdimensionaler Festkörper-^{13}C-NMR. PhD thesis Johannes Gutenberg-Universität Mainz (1994)

86. Suter U. W., Flory P. J. (1975) Macromolecules 8:765–776

87. Mays J. W., Fetters L. J. (1989) Macromolecules 22:921–926

88. Flory P. J. (1984) Pure Appl. Chem. 56:305–312

89. Theodorou D. N., Suter U. W. (1985) Macromolecules 18:1467–1478

90. Bartlett R. J., Stanton J. F. (1994) Applications of post-hartree-fock methods: A tutorial. In : Lipkowitz K. B., Boyd D. B. (eds) Reviews in Computational Chemistry. Vol 5. VCH, Weinheim, New York, p 65

91. Mansfield K. F., Theodorou D. N. (1991) Macromolecules 24:6283–6294

92. Komoroski R. (ed). (1986) High Resolution NMR Spectroscopy of Synthetic Polymers in Bulk. VCH Publishers, New York

93. Tanaka T., Chatani Y., Tadokoro H. (1974) J. Polym. Sci. Polym. Phys. Ed. 12:515–531

94. Schilling F. C., Tonelli A. E. (1980) Macromolecules 13:270–275

95. Bax A., Szeverenyi N. M., Maciel G. E. (1983) J. Magn. Res. 55:494–497

96. Maciel G. E., Szeverenyi N. M., Sardashti M. (1985) J. Magn. Res. 64:365–374

97. Terao T., Fujii T., Onodera T., Saika A. (1984) Chem. Phys. Lett. 107:145–148

98. Schönhofer R. Trennung verschiedener Kernspinwechselwirkungen in anorganischen Festkörpern. Master's thesis Johannes Gutenberg-Universität Mainz (1994)

99. Eckert H (1992) Progr. NMR. Spectr. 24:159–293

100. Krevelen D. W. van, Hoftyzer P. J. (1976) Properties of Polymers. Elsevier, Amsterdam

101. Brandrup J., Immergut E. H. (eds). (1989) Polymer Handbook. John Wiley & Sons, New York

102. Boyer R. F. (1975) J. Polym. Sci. Macromol. Symposia 50:189–242

103. Akiyama S., Komoto T., Ando I. (1990) J. Polym. Sci. Polym. Phys. Ed. 28:587–594

104. Ando I., Yamonobe T., Asekura T. (1990) Progr. NMR Spectr. 22:349–400

105. Abe A., Jernigan R. L., Flory P. J. (1966) J. Am. Chem. Soc. 88:631–639

106. Yoon D. Y., Flory P. J. (1976) Macromolecules 9:294–299

107. Yoon D. Y., Smith G. D., Matsuda T. (1993) J. Chem. Phys. 98:10037–10043

108. Smith G. D., Yoon D. Y. (1994) J. Chem. Phys. 100:649–658

109. Boyd R. H., Breitling S. M. (1972) Macromolecules 5:1–7

110. DeBolt L. C., Suter U. W. (1987) Macromolecules 20:1424–1425. Erratum ad [23]

111. Büchner S., Heuer A., Spiess H. W. Work in progress

112. Pham Q. T., Petiard R., Waton H., Llauro-Darricades Marie-France. (1991) Proton and Carbon NMR Spectra of Polymers. Penton Press, London

113. Schindler M., Kutzelnigg W. (1983) J. Am. Chem. Soc. 105:1360–1370

114. Hobson R. J., Windle A. H. (1993) POLYMER 34:3582–3596

115. Hobson R. J., Windle A. H. (1993) Makromol. Chem. Theory Simul. 2:257–262

116. Williams A. D., Flory P. J. (1969) J. Am. Chem. Soc. 91:3111–3118

117. Williams A. D., Flory P. J. (1969) J. Am. Chem. Soc. 91:3118–3121

118. Mark J. E. (1972) J. Chem. Phys. 56:451–458

119. Smith G. D., Jaffe R. L., Yoon D. Y. (1993) Macromolecules 26:298–304

120. Sundararajan P. R., Flory P. J. (1974) J. Am. Chem. Soc. 96:5025–5031

121. Sundararajan P. R. (1986) Macromolecules 19:415–421

122. Yoon D. Y., Flory P. J. (1975) POLYMER 16:645–648

123. Yoon D. Y., Flory P. J. (1976) Macromolecules 9:299–303

124. Vacatello M., Yoon D. Y., Flory P. J. (1990) Macromolecules 23:1993–1999

125. Apel U. M., Hentschke R., Helfrich J. (1994) Macromolecules 28:1778–1785

126. Miller R. L., Boyer R. F. (1984) J. Polym. Sci. Polym. Phys. Ed. 22:2021–2041

127. Miller R. L., Boyer R. F. (1984) J. Polym. Sci. Polym. Phys. Ed. 22:2043–2050

128. Kulik A. S., Radloff D., Spiess H. W. (1994) Macromolecules 27:3111–3113

129. Kulik A. S., Spiess H. W. (1994) Macromol. Chem. Phys. 195:1755–1762

130. Schmidt-Rohr K., Kulik A. S., Beckham H. W., Ohlemacher A., Pawelzik U., Boeffel C., Spiess H. W. (1994) Macromolecules 27:4733–4745

131. Kulik A. S., Beckham H. W., Schmidt-Rohr K., Radloff D., Pawelzik U., Boeffel C., Spiess H. W. (1994) Macromolecules 27:4746–4754

132. Tonelli A. E. (1991) Macromolecules 24:3065–3068

133. Tonelli A. E., Schilling F. C., Starnes Jr. W. H., Shepherd L., Plitz I. M. (1979) Macromolecules 12:78–83

Author Index Volumes 21–35

Printing: Saladruck, Berlin
Binding: Buchbinderei Lüderitz & Bauer, Berlin